MOVING THE OBELISKS

A chapter in engineering history in which the Vatican obelisk in Rome in 1586 was moved by muscle power, and a study of more recent similar moves.

by BERN DIBNER

The M.I.T. Press
Cambridge, Massachusetts, and London, England

First M.I.T. Press edition 1970

SBN 262 54010 X (paperback)

Library of Congress catalog card number: 79-107990

To the memory of

BENJAMIN HALPRIN

Civil Engineer, Captain of Ordnance

Legion of Merit

Natalis Bonifacius. Sibenicen. Dalmatinus Incidebat. Romæ. 1589. Cum Priuilegio Summi Pont.^{cis}

MOVING THE OBELISKS

N the middle of Central Park in New York stands a tall stone on a quiet, wooded knoll. It has stood here for 70 years and in this time has witnessed the neighboring streets swell in activity from suburban quiet into the busiest thorofares of all time. It has watched great buildings grow from the ground and it has been dwarfed by their eminence and bulk. It can, in all truth, say: "I have witnessed this great change in only one-fiftieth of my existence, for in my youth in Egypt I have had Moses look upon my face, and Joseph has paused within my shadow. I have seen a great city, as great as yours, burn and disappear and I have stood near the sea for 2000 years to witness another great city blossom and die. Be not proud, for I shall exist when all this brick and steel about me has crumbled into dust!"

This stone and others like it have been quarried, cut, engraved and erected by men for reasons of interest to us. They have been chiseled, raised, lowered and moved again by methods revealing to our engineers. Let him who can pause in his busy day to see what others have done, read further.

(Left): Frontispiece of Fontana's book on the transportation of the Vatican obelisk, published in Rome in 1590. This engraving shows the engineer-architect at age 46, shortly after his triumph. He holds a model of the obelisk and wears the insignia of his knighthood.

HE OBELISK was one of the three forms of monuments that evolved during the long history of the Egyptian state and religion, the other two being the sphinx and the pyramid. The form of the obelisk changed comparatively little in the two thousand years in which it was cut and mounted. Being monolithic in construction and tall and attractive in form, it still was not too massive to be moved by those who coveted its possession. The 200 to 500 ton shaft therefore became an item to be marked by the victorious invader and a challenge to his engineer to remove such a handsome trophy from Egypt and to have it set up in some distant land. That is why these superbly carved granite shafts, intended for some Nile necropolis or to adorn some temple, can be seen today, after three or four thousand years, as a trophy in Rome, or as a gift in London, Paris or New York. These ancient monuments are symbols not only of a religion no longer practiced, inscribed in a language unknown for the last thousand years, but also a challenge to the ability and ingenuity of the mason and the engineer right up to our own day. It must be noted that altho there are more than two score of obelisks and fragments still in existence* as evidence of the original craftsmanship, it still is a subject of high speculation as to how these great stones were shaped, moved and erected by men who had only the most primitive mechanical aids and no mechanical power as we know it today. Within the last century, three obelisks were moved from Egypt to the north and the west. These were one from Luxor in Egypt to Paris in 1836, one from Alexandria to London in 1877 and one from Alexandria to New York in 1880. Even tho steam engines, hydraulic jacks, steel cables and steel structural members were at their disposal, the mere moving and erecting of each of these obelisks became, to modern engineers, a major project of international interest and concern. Not only did the Egyptian engineers not have such modern aids but the cutting and finishing of the hard granite, its transportation over hundreds of miles, and its erecting, were accomplished by these ancients with a modesty that has kept such deeds from being adequately recorded. Whereas there exist thousands of sculptures, bas-reliefs, gems, paintings, papyri and models of the religious, regal and domestic life of the Egyptians, their advanced technology is illustrated by extremely few known examples. We must therefore reconstruct their tools and methods from the results they achieved.

Excluding the moving and erecting of the three obelisks that were transported in the last century, we are fortunate to have clear records of the mechanics used in the moving and erection of the Vatican obelisk in 1586. Unlike the former, in which modern mechanical aids were used, the obelisk standing today in the Piazza of St. Peter's, Rome, was moved by means that must have, in some measure, resembled those used by the Roman engineers, if not by the Egyptians themselves.

**For partial listing see page 59. For fuller listing see* GORRINGE *page 145 and* MOLDENKE *page 9. See also Bibliography on page 61.*

Egyptian Engineering

IN THE SPAN of 3500 years in which Egyptian civilization flourished, it was able to develop and support masons and engineers having highly developed mechanical skills and very high standards of perfection. There are today graven in stone or written upon linen and

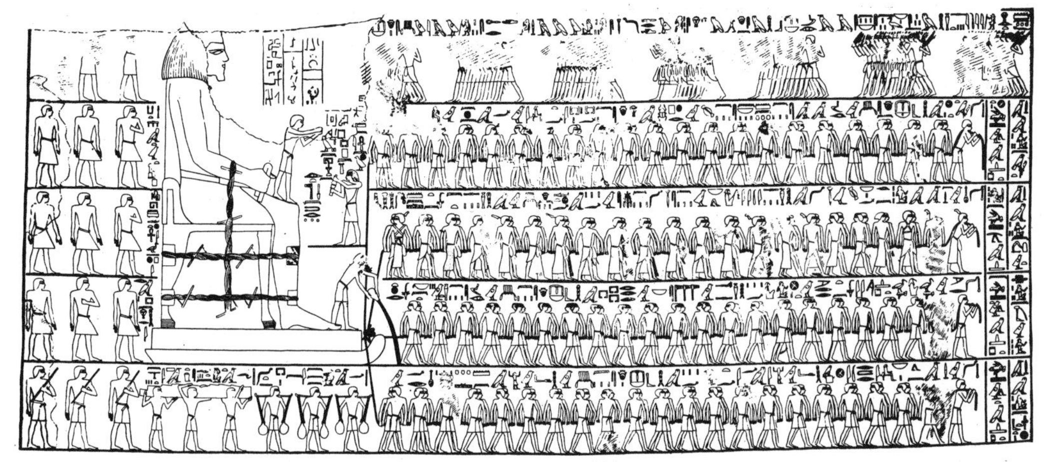

FROM NEWBERRY, EL BERSHEH, PART 1, PLATE XV, EGYPTIAN EXPLOR. SOC., 1893

Dragging the 132-ton colossus on a sledge required the pull of about 2000 men. The man on the knee of the colossus gave the signal for the concerted heave. The man at the toe poured oil before the sledge runners.

papyri, records of astronomical and mathematical computations, chemical, pharmaceutical and technological processes that must have involved guild practices and the maintenance of long records. Such records extend as far back as the Third Dynasty, or nearly 2800 years B.C., some time after Cheops had built the great pyramids in 3700 B.C. Later the conqueror and builder, Rameses II (19th Dynasty, 1400 B.C.), had massive statues of himself erected which, when completed, weighed about 1000 tons each. The one standing in the Memmonium at Thebes weighed 900 tons and was transported an overland distance of 138 miles. The pharaoh, Amenhotep III, similarly had carved colossal statues weighing 800 and 1000 tons that were erected 3500 years ago at Thebes. At Tanis in lower Egypt there was found the remains of a statue of Rameses so great that the large toe was the size of a man's body. When complete it must have stood 92 feet high and weighed more than 1000 tons. The remainder of this massive statue had been carved up by Orsoken II in 900 B.C. for the construction of a temple pylon.

So prolific were the Egyptians in the cutting and placing of their monuments, that many are still to be seen there even tho 2000 statues, it has been written*, were carried off to Babylon, while dozens of obelisks have been taken to Nineveh, Constantinople and Italy, in addition to those mentioned earlier. The pharaohs who built of hard and durable granite in such great bulk and quantity did so knowing how destructive later man might be, for were it not for this vandalism, nearly all the monuments of Egypt would be standing there today. The city of Memphis was taken apart stone by stone to build Cairo, leaving only the statue of Rameses standing in the sandy waste.

Egyptian engineers performed their great tasks depending upon such simple mechanical means as the lever, roller, inclined plane and wedge—and much muscle power. They may have used the capstan but seemingly did not know the compound pulley. What they lacked in mechanical advantages they made up in great amounts of disciplined and coordinated manpower. The emphasis seemed more on the human muscles than that of animals because the human muscles could be better coordinated

*BARBER, *page 4*.

and controlled. To these basic mechanical aids, Roman engineers later added the power of torsion from a twisted rope, shrinkage from wetting and the elasticity of the spring. The brace-and-bit and the whip-drill for boring holes are shown on tomb paintings, tho the bit was probably the spoon and not the screw type. The inclined plane or ramp was one of the most frequently used helps, for at Gizeh the remains of two inclined planes still exist. Sun-dried brick made of Nile mud (as described in Biblical accounts), was readily available for the building up of the great volumes needed to construct the causeways and ramps as required in the construction process.

Egypt, at its height, was the most populated country in all history, exceeding in density per square mile even our own records. Manpower was therefore in great abundance and from the social character of Egypt's institutions, readily available. Her 11,500 square miles held a population of 8,000,000, or a population density of 700 per square mile, approached today only by Holland and Belgium. China's population density is 120, and India's, 107. Survival was relatively easy in a warm land abundant in dates and durra, a form of millet or guinea-corn. During the season that the Nile flooded, the unemployed agricultural population was put to work on various construction projects including the tombs and personal monuments of the pharaohs. The number of men working on the great pyramid of Cheops (3730 B.C.) required the labor of 100,000 men according to Herodotus (366,000 according to Diodorus)*, working 20 years to erect this, the largest and oldest of man's monuments. The outer and chamber stones of this pyramid were carved at Syene, 560 miles up the Nile, and some weighed up to 60 tons. The gallery and outer stone faces were polished like glass and were joined almost invisibly; the fill was of limestone blocks. The inclined stone-topped road leading to the pyramid required as much manpower to build as the pyramid itself, for this road required about four times the volume in brick and stone chips. The completed pyramid of Cheops weighed 6,740,000 tons. Commander Barber* has calculated that if this great monument were to be budged by the human muscle power that went into its construction it would require the simultaneous heave of over 100 million men. He is convinced that only human labor went into the construction of the pyramid. In the 1200s A.D. this monument was recorded as being in perfect condition. Since then, its granite cover was stripped to provide stones for the building of Cairo.

The indications are that oxen were used for dragging stones of comparatively small size. Horses were introduced only after the Eighteenth Dynasty, the period of the Semitic Shepherd Kings called the Hyksos, a period we have learnt about from the Biblical accounts. It was the muscle power of man rather than of draft animals that provided the greatest source of power for moving these large stones and monuments that have become so characteristic of the Egyptian social order.

The Great Wall of China was built in the short space of only ten years and represents the greatest single engineering task man has ever completed. It was finished in 204 B.C., is 1500 miles long (it would stretch from New York to beyond Houston, Texas), is 30 feet to 40 feet high, 25 feet thick, and has towers at frequent intervals. It is built of bricks weighing approxi-

*Barber, *page 29.*

**See Bibliography on page 61.*

mately 60 pounds, with rubble fill. It crosses mountainous territory and at one peak reaches an elevation of 5,000 feet.

In Hyderabad there was hewn the remarkable temple or Kylas at Ellora. This is a veritable cathedral cut out of solid rock 365 feet long, 192 feet wide and 96 feet high, with chambers hollowed out and walls decorated in stone sculpture, both inside and out. When one ponders at the chipping required in the shaping of even a simple stone, the man-hours that must have been dedicated to the forming of these great monuments is indeed awesome and proves an incontestable fact, that there were the man-hours to so give.

In the complex Egyptian religion, pantheistic and given to animal worship, the statue of the pharaoh was equivalent to that of the greatest gods. To insure such glory for the eternity that formed one of the elements of his religion, the pharaoh proceeded to immortalize himself and his station by the most permanent means he could think of—a granite monument to himself—making sure that all his titles were engraved upon it. It is therefore natural that the architectural form into which the symbols of glory evolved should have such stable forms as the pyramid, colossus and pylon. The obelisk form is less easily understood and there is no agreement among authorities as to why it assumed

COURTESY NEW YORK PUBLIC LIBRARY

One of the pair of obelisks moved by the Romans from Heliopolis to Alexandria where it stood for 2000 years. The fallen one nearby was taken to London in 1877, the one standing was lowered and shipped to New York two years later.

its general shape. Some believe it to be an evolution of a phallic symbol, others the representation of a ray of sunlight, thereby dedicating the obelisk to Ra, god of the sun. The Romans believed it to be a gnomon, or sundial, and tried to put it to that use. It appears that obelisks, dedicated to the rising sun, were erected on the east bank of the Nile, but pyramids, the mausolea of the kings, on the west bank.

Of the hundreds of obelisks that once glistened under the Egyptian sun, the number has been reduced to a grand total of less than 50. Although some may have fallen due to the ravages of earthquake and of time, it seems that the vast majority have disappeared because of the hunt for treasure, the religious fanaticism of competing cults and those who altered monuments to emphasize their own greater glory. Of the few that have remained, even fewer are of authentic Egyptian origin or in good condition. The greatest cluster of obelisks existed in the ancient city of Heliopolis, near modern Cairo, put there mainly under Usortesen I of the 12th Dynasty, and followed by his successors for a thousand years. It was in the shadow of the "upright stone" as mentioned in Jeremiah (xliii:13) that Joseph was wedded to the daughter of the high priest. Near the obelisks, in addition to the university, there stood the Temple of the Sun, rivaled only by the shrine of Ptah at Memphis, the most ancient of all Egyptian temples, a temple which at one time was served by 13,000 priests and servants. Heliopolis is referred to in the Bible by the name *On* and in Egytian by *Annu*, meaning *obelisks*. At its university studied Moses (Acts, vii:22), and later its streets were trodden by Solon, lawgiver of Athens, and by Thales of Miletus who first described an electrical phenomenon. Its library was the largest in the ancient world; it was transferred to Alexandria by Ptolemy Soter in 305 B.C. At the time of the Greek kings, the Ptolemys, who ruled Egypt after it was conquered by Alexander*, Heliopolis was already in ruins. One of the ancient travelers records the "innumerable" obelisks he saw in Heliopolis. Of these a modern census can trace eight. Three

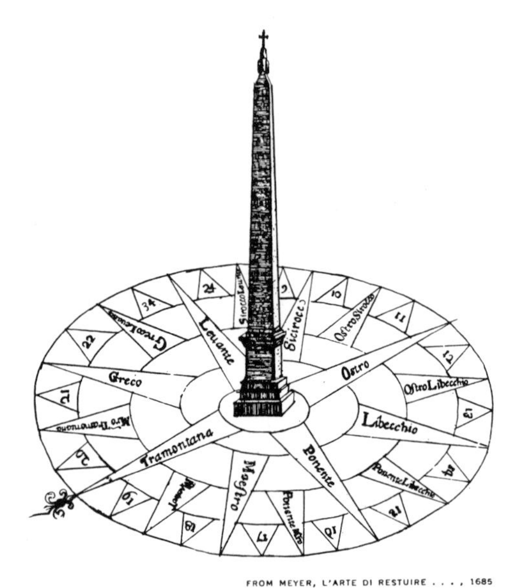

FROM MEYER, L'ARTE DI RESTUIRE . . . , 1685

Use of the obelisk as a sun-dial was tried by the ancient Romans, and was again proposed after its restoration as an archaeological and architectural element in modern times.

are in Rome, one in Florence, two in Constantinople, one in New York and one in London. It was Cambyses II, son of Cyrus and King of Persia who, in 525-521 B.C., largely destroyed Heliopolis. Many of its great stones were later used to build up the comparatively modern city of Cairo. An earlier Assyrian king, Asshurbanipal moved two obelisks to Nineveh as trophies and thereby set a pattern followed by many subsequent conquerors. Nine obelisks are still in Egypt, nine of Egyptian origin are in Rome, ten are fallen or exist in broken form.

**Alexander (356-323 B.C.), in turn, founded what later became the center of new Greek culture, the city of Alexandria. Here Euclid founded the school of mathematics and here Archimedes studied.*

The Obelisk

The shape of the obelisk hardly requires definition. It always possesses three characteristic elements—the square, tapering shaft; the pyramidion of 60° taper at its top; the monolithic form. Its shape seems to have evolved from the earlier stela form (shaped like a milestone). The Washington Monument is not a true obelisk since it is formed of many individual stones and is hollow. The word obelisk is derived from the Greek *obeliskos*, meaning a skewer or roasting spit. In their most frequent use, obelisks stood in pairs on either side of temple gates, usually behind a similarly-placed pair of sphinxes but before a pair of colossi standing in front of the actual temple entrance. The obelisks were also used as sepulchral markers in a necropolis or sacred field of the dead. The general forms are square sections tapering upward to a height ten times the side dimension. They usually rested on a cubical pedestal which, in turn, was set on a base forming one or two steps. The pyramidal top was usually sheathed in bright metal, such as gold, gilt bronze or electrum (four parts gold, one part silver) so that its facets reflected the moving sun like a beacon or heliograph upon the landscape, thus making it seem like a source of sunlight and therefore the throne of the sun deity. The faces of the obelisk were used for inscriptions praising the glory of the pharaoh. Sometimes the inscriptions would be effaced or altered by succeeding kings and sometimes they would be added to. The obelisk in New York contains inscriptions of three such kings, Thothmes III, Rameses II and Orsoken I.

All Egyptian obelisks came from a common quarry, located in the Lybian Mountains at Syene (Assuan in Arabic) just below the First Cataract on the east bank of the Nile, about 700 miles from Alexandria and 560 miles from Heliopolis. Here in the hillsides there exists a level of reddish granite of great depth, fine texture and free from seams and flaws. This amphibole-granite or syenite has, in addition, a hardness and lustre that combine all the qualities of beauty and permanence that satisfied the requirements for the immortality the Egyptian kings sought. From this scene came not only all the obelisks but also many other stones out of which were formed the massive monuments of Egypt. Not much light is given us as to the methods the ancient masons used in cutting and polishing their stones. However, some evidence

FROM AGRIPPA, NUOVE INVENTIONI . . . , 1595

A book proposing the method for moving the Vatican obelisk, written by Camillo Agrippa, architect, navigator and cavalier, gave the impetus that changed a century-old dream into a reality.

still exists at this quarry at Syene, including an obelisk 95 feet long that was in the process of being cut from the rock and was abandoned after the three upper faces were shaped. The fourth, or bottom face, is still a part of the mother rock.

The most agreed-upon method of quarrying stone at Syene consisted of marking the line of cleavage by cutting a channel two inches wide by two inches deep along the line of fracture. Along this groove there was drilled a line of holes about three inches in diameter, six inches deep and spaced about eighteen inches apart. Some authorities say that wooden plugs were driven into these holes and that water was poured upon these plugs and into the channels. This caused the plugs to expand and burst the rock along the grooves. Another authority* claims that fires were built over the two inch by two inch groove and were kept burning for a long time so that the rock would be deeply and uniformly heated. At a signal from the foreman the burning embers would be quickly swept away and the workmen, placed on either side of the groove, would then quickly pour cold water into the groove. The sudden contraction would cause a sharp fracture in the rock along the groove line.

There are some who have tried to determine how the quarrying operations were carried out by observing what practices were used in the same level of technological development in other parts of the world. It has thus been reported by a Col. Wilks, R.A., who, in 1821, watched such quarrying operations in India.** He observed that after marking the line of cleavage, workmen would chisel a two-inch by two-inch groove along the line of cut. Along this groove there would be drilled a series of holes about eighteen inches apart. At each of these holes a man with a chisel and iron mallet was set and upon a prepared signal each man struck his chisel in a wave sequence from right to left

*Sir J. F. Herschell as reported by BARBER, page 71.
**BARBER, page 72.

or from left to right along the line. This required repetition over two or three days but would finally end in a neat fracture of the rock.

The exact composition of the tools used by the Egyptians in their process of quarrying, cutting and polishing remains unknown because, with one exception, such tools have not been found or described. Wilkinson* reports having found a bronze chisel having 9.9% tin among the limestone chips in a tomb at Thebes. This had an edge 7/8 inch wide which, altho still quite sharp, was so soft that it turned when it was struck against stone. Shape and composition can only be surmised from the work which these tools accomplished. Wall paintings do show the use of hammer and chisel but whether these are of bronze or iron cannot be determined. It is presumed that finishing operations were done with saws but no picture of these is extant. It is also presumed by Petrie** that the tools must have been bronze with diamond teeth and they were fashioned to saw, drill and polish the hard granite. The soil of Egypt is generally nitrous and would therefore be destructive to iron over a long period of time. For this reason, whatever iron tools may have been used have disappeared. Iron mines are known to have existed in Egypt and in Phoenicia. It is also argued, tho not proved, that the cutting of hieroglyphics and the polishing were done at the quarry. This might have been intended to reduce the weight to be transported and also to show up any imperfections in the stone before it was moved. In the polishing process it cannot be ascertained whether emery or diamond dust was used. The French engineer, Le Bas, who was in charge of the moving of the Paris obelisk from Luxor and who studied the mechanics of the obelisk in great detail,

*Reported in BARBER, page 75.
**Reported in BARBER, page 76.

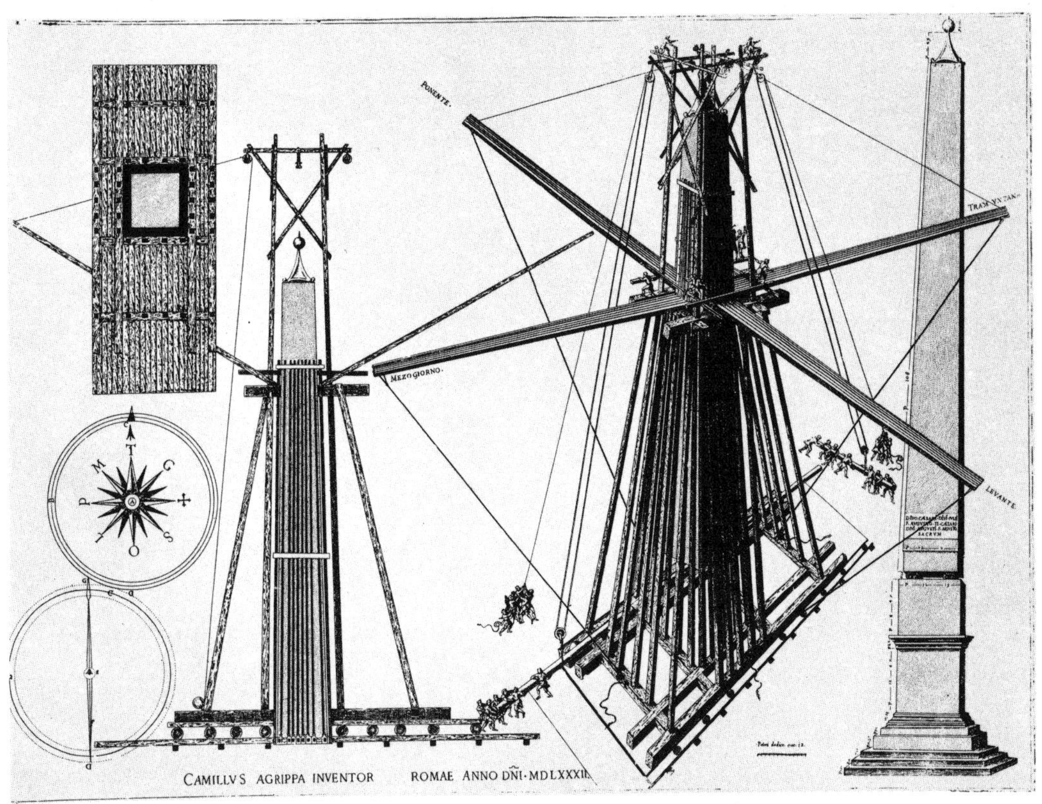

FROM AGRIPPA, TRATTATO DI TRASPORTAR LA GUGLIA, 1583

The means proposed by Agrippa for lifting and moving the ancient obelisk consisted of giant shores and levers set upon a carriage moving over rollers. A more practical arrangement was later adopted.

concluded that to obtain such polish all that was required was sandstone, pumice-stone and much time. Similar operations in India used circular stones hollowed at their base to hold the polishing agent consisting of a mixture of corundum and beeswax. These stones can be twirled by two men using two sticks with connecting cords about the stone's middle. This will give the rotating stone, shaped like an hourglass, a constantly circular rotary motion.

The fine finish of the hieroglyphic inscriptions on the hard granite indicates a definite hardness of tools that must have been the equivalent of our own tool steel. Examination of the intaglio cut seems to indicate the use of pointed rather than spade chisels. It has also been proposed that in these operations the drills had diamonds set into a circularly-cut head which, when twirled by the bow and cord, produced an effective cutting tool.

The straining for perfection in the cutting and polishing operations was so high that modern analysts using modern measuring instruments marvel at the fine tolerances attained. The obelisks still standing in their original positions in Egypt have retained a polish nearly as high as when they were first erected.

Wonder as we will at the difficulties that had to be overcome in the quarrying, cutting and polishing of these great stones, we stand in greater awe at the task performed in transporting and erecting them. Evidence describing these processes is most meagre. In general, it is presumed that the stone was dragged from the quarry to a dry-dock prepared for it on the dry Nile embankment. There the stone would be dragged and skidded onto a float or barge set within this dry-dock. When the rising of the waters in the rainy season reached a predetermined level, the wall separating the dry-dock from the river would be pierced and the river would enter the basin and float the barge and its obelisk. The barge and its load were then floated down the Nile for many hundreds of miles and at its landing site the process would be reversed, until it arrived at its pedestal. The method used for the translation of the obelisk from its horizontal to a vertical position is completely unknown to us.

It was found on examining the obelisk still uncut in its quarry bed at Assuan that its lower side was pierced by a number of parallel tunnels. It is therefore presumed that these tunnels were to be fitted with timbers which, when the lower face was completely cut away, would form cross-members of a sledge upon which the obelisk would be drawn to the river's edge. Good evidence of this exists in a single wall painting discovered about a century ago on the walls of a tomb at El Bersheh, shown here on page 9. This painting (partly damaged) and its hieroglyphic text describe the moving of a colossus on a sledge. It was made during the reign of Usortesen II, in the period of the 12th Dynasty (2000 B.C.).

From the picture and text, Barber* has calculated that the colossus weighed 132 tons and that it would require about 2000 men to drag this weight. The picture itself shows only 172 men at the dragging ropes but this is only a token representation of manpower. The use of men instead of oxen, as has already been indicated, was to permit a concerted pull at a signal by the clapping of the hands of a man perched upon the knee of the colossus and repeated by a drummer below him. Of particular significance is the man shown pouring oil in front of the runners of the sledge. No better example of illustration of Egyptian engineering exists.

*BARBER, *page 19.*

To pull an obelisk with such equipment, it has been estimated that some 50,000 men, arranged in two double files, would be required. Similarly, other limiting factors that determine the maximum size of the block of stone to be dragged were: rope size and number, the distance of control by the foremen and the mobility of the pullers as a unit. These limited the mass so hauled to about 1200 tons. The Nile became the main highway for such heavy burdens. Laterally from the Nile highway these great monuments were dragged by thousands of men for great distances, as attested to by inscriptions dedicated to those who managed the moving and those who died at their labors.

The loaded barge would be drawn up the river in tow by groups of some 30 boats manned by about 960 rowers. Dragging-ropes for these operations were made from the fiber of the date palm, a fiber possessing excellent qualities for making strong rope. Wall paintings show its many applications in haulage operations, as shrouds and rigging of boats and as harness.

A wall painting that shows two obelisks lashed to the deck of a barge was found at Deir el Bahri*. It is but a fragment of an extended wall painting destroyed by Coptic Christians, who obliterated the symbols of their pagan predecessors.

Because Egyptian theology regarded life on earth as simply one in a series of long transi-

*BARBER *plate XVII.*

tions, the element of time, a lifetime or historical time, was of no great importance and therefore Egyptian chronology is either non-existent or most confusing. This is true with regard to obelisks—with one exception. At Karnak there still stands an obelisk, the easterly one, the mate of which is now in Paris. On the base of this shaft is engraved the legend that this stone was cut, engraved, polished and moved in a period of seven months. From the month in which it was finished, it is presumed that this project was hurried in order to get the stone on its dry-dock preparatory to having it floated by the rising waters of the Nile.

The great era of construction of the granite and limestone monuments covered a period of some 3000 years of Egyptian history, ruled over by thirty dynasties of pharaohs. This was followed by three centuries of rule after the conquest of Egypt by Alexander. It must be remembered that some 2500 years of Egyptian development lay behind the Greek conquest, a period longer than from the days of Caesar to today. The elaborate temples, pyramids and busy cities bore accumulated marks of century after century of construction and accomplishment—accomplishments that are of great wonder even today. Egypt was the natural home of the stone mason, there being no construction timber available, whereas the cliffs were ever at hand. With an abundance of stone and manpower, a culture that provided such massive monuments and highly precise workmanship had thus evolved.

The Roman Period

The well disciplined, well organized Roman administrators, backed by the military power of the Caesars, conquered Egypt in 30 B.C. and made her a Roman colony. There then followed a period well covered in history and drama. To Roman engineers, experienced in the construction of roads, forts, colossea and aqueducts, the sight of the work of Egyptian engineers and architects must have been admirable and challenging. It therefore follows that at least fifteen of Egypt's obelisks should be moved to Roman territory. Of these, twelve, now in various conditions of completeness, still stand in Italy. The Romans not only moved the Egyptian monuments to Italy, they even imitated their form, altho with inferior workmanship and in inferior materials. To facilitate their operations, the Romans used mechanisms unknown to Egyptians, in particular, the compound pulley. It is recorded that they dragged their burdens by using a *chalmalcus* which consisted of a cradle-like sledge drawn on greased runways by means of tackle and capstans.

The records of Roman operations are almost as unknown as those of the Egyptians. Roman engineers undoubtedly were accustomed to the use of timber as structural members, unlike the Egyptians who depended so much on stone. In addition, Vitruvius, an architect and military engineer to Augustus, living in the first century B.C., described a two-legged shears which supported tackles. In addition, the motive power came from men working in a squirrel-cage, thereby giving the mechanism the wheel-and-axle mechanical advantage. It was probably this device, called a *polyspaston*, that was used by the Romans in the erecting of the Lateran obelisk in Rome at the time of Constantine, 345 A.D. Further, the Romans, being a Mediterranean power, were more advanced nautically than the Egyptians who distrusted salt water and who looked to the Nile for its maritime existence. This resulted in more powerful boats and barges. The boat that carried the Vatican obelisk across the Mediterranean had 300 rowers and, in addition to the obelisk, it carried a ballast of 1000 tons of grain, placed in sacks and fitted around the obelisk to keep it from shifting when the vessel rolled.

Using timber, a heavy framework was erected over the pedestal to receive the obelisk. Tackles were then attached to bands of rope placed at frequent intervals along the obelisk's length and to this were lashed the ropes from the tackles. The obelisk was then raised vertically and lowered onto four crab-shaped supports placed at each of the corners of the pedestal. These crabs were of bronze, about 16 inches in diameter, and had a shank extending downward into the square hole cut into the pedestal, and another extending upward that fitted into the obelisk shaft. These shanks were square and in a section that expanded outward. When in place, lead was poured around these dovetail-shaped shanks so that future vandals could not easily remove the bronze crabs. The crab form was chosen as the appropriate symbol because the obelisk was dedicated to the sun-god and the crab was the symbol of the sun-god Apollo. The Egyptians placed their obelisks directly upon the top base of the cubical pedestal—not an easy feat. The base, therefore, had to be absolutely level so that the apex of the obelisk would rest truly vertically over its base. The Romans, like the Persians, used crabs as astragals because these provided a space for slings between the obelisk foot and the top of the pedestal.

The obelisks, brought from Egypt and also copied in Italy and France, were used as items of adornment in the circuses and public squares of ancient Rome. There they witnessed the triumphs and tragedies of the Eternal City. With the passing centuries the sun began to set upon the Roman Empire and in 410 A.D. Aleric and the Visigoths were storming the gates of the city. For a thousand years thereafter the city decayed, grass grew in the market places, carefully laid stones were toppled one upon the other, and wolves roamed the empty places.

In the 1300s, the Italian Renaissance began and the pens of Dante, Petrarch and Boccaccio again gave voice to the subdued people. Men once again took pride in their family, their faith and their community. Buildings of grace and beauty began to appear and Florence, Pisa and Milan again rang with the mason's hammer. The engineer was once again called upon to build and to reconstruct.

In this millennium the Papacy had gone thru a series of changes in the concept of its role, moving all the way from pure theocracy to that of being one more political state. Pope Leo III dreamt of a Holy Roman Empire to embrace a perfect union of church and state to cover all of Western Europe. In the year 800 in the old basilica of St. Peter's in Rome, Leo crowned Charlemagne with the iron crown of Emperor, but such perfect equality did not remain long in balance, and wars between crown and mitre swept back and forth across Europe. The Reformation, by dividing Christendom, helped split the Holy Roman Empire and the Holy See was further weakened when Clement V, a Frenchman, elected to the papacy, decided to transfer its seat from Rome to Avignon. The Italian church dignitaries would not abide by such a decision so that dissension, lasting 120 years, followed, finally resulting in the establishment of the center of church authority once again in Rome in 1377, and this was followed by the twin rule of Pope and anti-Pope for some forty years. This stormy period subsided with the election of Nicholas V whose high ideals were markedly different from his more material predecessors. Rome had grown rapidly from a city of some 50,000 in 1560 to twice that number in 1572, at a time when Paris boasted a population of a third of a million.

FROM FONTANA, TRASPORTATIONE DELL' OBELISCO, 1590

In his famous book describing the successful transportation of the Vatican obelisk in 1586, Fontana showed the partly-buried ancient stone and a few of the unused plans proposed for moving it. At the top is shown his successful tower. Seen also is the old sacristy and the unfinished walls of St. Peter's.

Obelisks in Rome

Rome can boast having more obelisks than any other city, a dozen, in various sizes and conditions of preservation. Among these can be counted the heaviest, the best preserved and the tallest.

This obelisk, the tallest known to exist, was brought by Constantine from Heliopolis. He had it erected in the Circus Maximus and from here Pope Sixtus V had it moved to the Piazza di San Giovanni Laterano. The Roman Emperors Hadrian and Domitian copied the Egyptian obelisks and had these, with Latin inscriptions, erected in Rome. Augustus transported two of the Egyptian obelisks to Rome and left two more in Alexandria, these being the two that were later moved to New York and London. Domitian moved two more from Egypt to Rome.

Pliny relates that the Emperor Augustus used the obelisk in the Campus Martius as a gnomon, or sundial. He had laid out a stone pavement and had the positions marked where the sun's shadow of the obelisk's point stood at high noon on the day of the winter solstice, and correspondingly where the length of the shadows shortened on succeeding days till it reached that of high summer. These positions were marked in bronze lines inserted in stone under the direction of Facundus Novus, the mathematician. To make the shadow of the obelisk tip sharp and discernible, he placed upon its summit a gilded bronze ball, an idea he got from observing the shadow of the human head. Pliny observed further that, in succeeding years, these observations did not agree with those first made; whether because of the change of the sun's course, a displacement of the earth's center, an earthquake or settling of the obelisk's foundations, he did not know.

With the resurgence of wealth and power to Rome, there was also restored the will to re-create this metropolis to something of its former glory. Successive popes sponsored the restoration of public buildings and the construction of new churches. In the late 1400s, Alexander VI brought Bramante (1444-1514) in as his chief architect and there, under the ambition and drive of Julius II and especially under the magnificence of Leo X, Rome began to take form as we know it today. Bramante, and later Michelangelo (1474-1563), began to erect the most magnificent cathedral in all Christendom. St. Peter's stood on the site of the ancient church erected in the year 334 during the reign of Constantine, the first Christian church erected in Rome. Over the tomb of St. Peter, Bramante planned a magnificent dome, and dome and church required a century and a half to complete.

The Vatican Obelisk

With only the main arches and walls of St. Peter's completed and the dome still unbuilt, the College of Cardinals met in 1585 to elect one from among them to become the new pontiff to succeed the deceased Gregorius XIII. The choice was Cardinal Montalto and he ascended the pontifical throne as Sixtus V at the age of 64. He, like Julius II, was destined to imprint his vigorous personality upon the face of Rome. It was his determination to continue the construction and decoration of St. Peter's into the most handsome structure in the world, and as part of this scheme it was proposed to include the majestic old obelisk that had been standing since Caesar's day on its pedestal behind the sacristy of the new St. Peter's. Alone among the obelisks of Rome, it had stood erect while the Circus of Nero crumbled from neglect and from the pillage of its stones to be used in the newer structures. Not one other of the many obelisks had so well withstood the ravages of vandalism and time, none other had remained upright or unbroken.

Prior to its being brought to Rome it was known that the Roman Emperor Caius Caligula (37-41 A.D.) and the Emperor Claudius (41-54 A.D.) had had it transported from Heliopolis to Rome and set on what became the Circus of Nero. It was the third of many to be so moved. It had no inscriptions on its faces prior to being brought to Rome so that its more ancient history could not be determined as accurately as those obelisks having hieroglyphic inscriptions. However, it was believed to have been cut at Syene in the 13th century B.C., during the reign of the pharaoh Noncoreo* in gratitude for the restoration of his sight, and similarly, according to Pliny, it is believed to have been raised to commemorate the deeds of Sesostris (Rameses II). On two of its bare faces, the east and the west, Caligula had ordered the following inscription to be engraved:

DIVO. CAES. DIVI. IVLII. F. AUGUSTO.
TI. CAES. DIVI. AUG. F. AUGUS. SACRUM.

After the sacking of Rome, the area around the Vatican shaft remained first as a public center and then as a place in which refuse was dumped. The ground around it gradually rose so that it completely covered base and pedestal, reaching right up to the base of the shaft itself. Further desecration consisted of removing ornamental bronzes at the foot of the shaft, leaving 24 holes as an indication that bronzes had at one time existed there. These had been placed there by the Romans and were reported removed by the Bourbon soldiers during their invasion of 1376. Similarly, the large gilt bronze ball that crowned the pyramidion had been shot at and dented by archers, probably of the same invading army.

This name does not appear in the lists of Egyptian Kings and is believed by Bunsen to be that of Meneptah (1322-1302 B.C.).

It was this proud symbol of the triumph of Christian Rome over pagan Rome that prompted many popes to plan the inclusion of the obelisk into the pattern of the new Rome. Its old position was awkward because it prevented proper framing of the back of St. Peter's. It required moving this enormous stone 275 yards from its old position to the new one in front of the cathedral, and it therefore became a topic of major interest among engineers as well as among prelates. It was forgotten that in its long history it had already been moved from the quarry at Assuan 560 miles to Heliopolis where it had stood for 1300 years. It was then taken down, embarked and brought up the Nile and across the Mediterranean to an Italian port. Here again it was landed and hauled to its Roman site. Not only it but others like it had come in a similar way. Yet how was it now to be moved simply and safely? Unknown were the practices of the Egyptian engineers, nor were there any helpful records of what the Romans did.

Three elements combined to transform a wish into a reality. The first of these was the appearance in the year 1583 of a small book written by one Camillo Agrippa (1535-1589), a Milanese, containing his proposals for transporting the giant stone. He stated that a problem under discussion for 50 years was to be resolved by his proposal and he included a copper engraving to show his mechanism.* This is shown on page 15. His scheme consisted, mainly, in encasing the obelisk in a sheath of oak timbers to protect its surface during transportation which was to be in an upright position (although he states that the ancient obelisks were moved prone). He then proposed that from a position about two-thirds

See AGRIPPA in Bibliography page 61.

FROM FONTANA, TRASPORTATIONE DELL' OBELISCO, 1590

The tower that Fontana built on either side of the obelisk to carry the tackle for lifting and lowering the monolith. To help raise the great weight, twin levers extend to the right, triple levers to the left into the pierced sacristy wall. At the top is shown the variety of blocks required, and the iron bands, bolts and wedges.

of the way up the shaft there were to be placed forty timbers slanting inward, twelve each on the two sides and eight each on both ends. He recommended that these timbers should rest on a frame and platform which, in turn, were to be set on eight rollers moving on a timber bed leading to the new site. In order to raise the shaft above its platform, his plan was to lift it vertically by a brace of eight oak levers operating on each of the four sides. These 32 timbers (the engraving shows only 24) were to be over 48 feet long, were to be equipped with grappling hooks on the weight end and operated by a system of rope and pulleys for lifting the stone itself. Agrippa had acquired a reputation as a philosopher, architect, fencing-master and navigator and the publication of this book started once again the discussion of the obelisk, a discussion revived and dropped since Nicholas V, active 150 years earlier.

The second item to bring reality to the obelisk plan was the ascension of Cardinal Montalto to the papacy. In his person were combined the ambition, firm will and great energy necessary to create a Rome fittingly reflective of his power and position. Sixtus V has been rightfully called "the creator of modern Rome" for, in the short period of five years in which he filled the pontificate, he had, in the popular tradition, "built five bridges and five fountains, erected five obelisks and left five millions in the treasury"*.

The third ingredient in this formula of success was the engineer and architect, Domenico Fontana (1543-1607). He was born in Mili (now Melide) on Lake Lugano in the Ticino canton of Switzerland. He thus springs from a nation famous for its engineers and also from a family that had given many sons to this constructive profession. He was the son of an architect Matteo Fontana who was active in Venice in the 1400s. In the 1500s three Fontana brothers, Giovanni, Domenico and Marsiglio, worked in Rome, often together as engineers and architects. The oldest of these, Giovanni (1540-1614), was retained as hydraulic and military engineer by four popes and by the King of Spain. Domenico was the second brother and he probably came to Rome to join Giovanni in 1563 and he there worked for Gregorius XIII, then for Sixtus V. Also, a nephew, Carlo Maderno, later built the new facade and portico of St. Peter's. Twenty years after Domenico Fontana's death, his nephew Giulio Cesare Fontana built a monument to his memory in the vestibule of the church of Santa Anna dei Lombardi on Monte Oliveto.

It may be considered that Agrippa was the catalyst that was responsible for bringing the obelisk problem up for final solution, because Sixtus and Fontana had known each other for many years. For Cardinal Felice Peretti de Montalto, Fontana had worked as the designer and constructor of the mausoleum of Pope Nicholas IV in 1574, and of several palace buildings. He was then building a small chapel of the Sacrament erected in the Santa Maria Maggiore, one of the four prime churches in Rome. This show of magnificence on the part of Cardinal Montalto seems to have displeased Pope Gregorius XIII, who thereupon suspended the Cardinal's income. Fontana was so wrapped up in his projects that he continued them at his own expense. This was not too unfortunate because on April 24, 1585, Montalto became Pope. Fontana was immediately repaid out of the papal treasury and was made papal architect. An immediate task assigned to him was to continue the work of Giacomo della Porta (1529-1604) in completing the construction of the dome of St. Peter's which Michel-

*Sarton, *page 4.*

angelo had left unfinished upon his death.

It is related that once while the then Cardinal Montalto and Fontana were crossing the Vatican area, the cardinal remarked that if he ever became pope the obelisk would not long remain in that place. Caligula had erected this obelisk in honor of the Emperors Augustus and Tiberius on the spina of the Circus Gianus which Nero had built on the Campus Vaticanus. The cardinal could well think back to the days when this old stone witnessed the martyrdom of many of the founders of his Church, among them the Apostle Peter, near the very foundations of the temple now being erected in his name. The moving of so important a monument from its poor location adjoining the sacristy would surely be a pious and worthy step, particularly if it would grace the center of the new piazza to be formed before the church dedicated to this martyr.

Agrippa's book had been in circulation for two years before Montalto had risen to the papacy. It was dedicated to the son of the previous pope and must have excited the attention of the ambitious cardinal. It is therefore not surprising that one of the first official acts of Sixtus V on assuming the papal throne was the appointment, on August 24, 1585, of a special commission to study the problem and to make a recommendation for the transportation of the obelisk, then considered as a problem both difficult and challenging. The commission met in the palace of its head, Pier Donato (Cardinal Cesis) and was composed of prominent men of Rome representing both the ecclesiastic and state functions of the papacy. In addition, there was representation from the administrative and legislative bodies as well as the departments of finance and public works. On this commission were such dignitaries as Cardinal Guastavillano, the papal chamberlain; Francesco, Cardinal Sforza; Ferdinando, Cardinal de Medici (later Grand Duke of Tuscany) and such functionaries as the Treasurer-General, Commissary-General and Commissioner of Roads. Secular Rome was represented by several Senators and Deputies — in all a commission endowed with a sufficiency of funds and authority and one committed to act. The dream of many popes was now to be realized.

The commission met immediately after its appointment and called in the engineers, architects and mathematicians whose opinions they solicited for the best solution of their problem. The subject excited the interest of learned men everywhere and at subsequent meetings of the council there appeared some 500 theoretical and practical construction men from all over Italy and from such remote places as Rhodes and Greece. Some brought with them plans, drawings and sketches. Others brought models and equipment and some just made oral statements. The proposals were as varied as the men proposing them. Some wanted the obelisk moved vertically, others horizontally, while those who would be different suggested moving it inclined at about 45°. All the prime mechanisms then known were recommended—levers, pulleys, jacks, cradles and rollers and, of course, many combinations of these. The commission listened to them all but was most impressed by the plan, model and presentation of the engineer Domenico Fontana. He presented his plan and demonstrated how the monolith could be lowered, moved and lifted by a combination of wooden tower and ropes and pulleys, using as a model a lead obelisk about two feet tall, and a wooden tower with ropes and tackle made to scale. The commission thereupon appointed Fontana engineer-in-chief in charge of the project.

We are thankful to him not only for performing a most workmanlike job in planning and executing a difficult physical and political task, but like the complete artisan he was, by leaving for posterity a record that, in format, type and engravings, makes one of the handsomest and most complete records of any engineering problem, a prize in the field of bibliography and engraving. His book *Della Trasportatione dell' Obelisco Vaticano* was printed in Rome in 1590. It is very rare* and has become a classic in engineering literature.

There were many contenders for the honor and it must be realized that among these, many were disappointed because they had not been selected. Palace intrigue, therefore, began to make itself felt and the commission that had made the selection began to feel this pressure. It hesitated in implementing its choice and then decided that Fontana was at this time too young (he was then 42 years old) to be fully responsible for so colossal an undertaking.

To make absolutely sure that all would go well, the commission thereupon appointed Bartolommeo Ammannati (1511-1589), a Florentine, as chief supervisor and della Porta as his assistant. Ammannati was regarded then as Italy's outstanding engineer-architect, and Fontana accepted this arrangement, at least outwardly, with some satisfaction for, said Fontana, while Ammannati was busy with major worries, he could concern himself with the technical problems involved. Whatever may have been the direct influence, before very long both Ammannati and della Porta were relieved of their positions, for Sixtus had taken seriously to heart Fontana's hint that "only the designer can best execute his own designs".** In his book, Fontana introduces the general subject of moving the obelisk by showing just a few of the variety of schemes (see page 19) proposed for moving and erecting the shaft. A look at these helps one to understand the technological status of the time. This view shows the obelisk in its old position as set by Caligula, the rounded sacristy and the unfinished dome. The main view shows how its base and pedestal had become buried under the newer ground level reaching up to the plinth. At its top is also shown the gilded bronze ball superimposed on it by the Romans.

Scheme B is that proposed by Camillo Agrippa. As has already been indicated, this proposal consists of lifting the obelisk by four sets of horizontal levers, a very ill-considered and impractical plan. Proposal C consists of rolling the vertical obelisk into a horizontal position by means of a half-wheel, somehow moving it and then re-erecting it. In view D, the obelisk is shown being raised and moved vertically by wedges. In E, Fontana shows a process which consists of lowering the obelisk into a reclining position with long screws and then moving it in a reclining position. Proposal F is another scheme using giant levers. In G, there is shown a ratcheted quadrant for lowering the obelisk a bit at a time and, finally, in H, the proposal was to raise the obelisk, shift it and re-set it, using a complex of screws. At the base can be seen a rose-of-the-winds which orients the operation with a similar indicator in several of the other illustrations; north is indicated by T (*tramontana*).

At the top, over all these rejected proposals, and borne on the wings of cherubs is A, Fontana's own winning proposal. This shows the combination of timber towers, guys, ropes and pulleys for lifting the obelisk and rollers for

**Library of Congress has no copy but copies are recorded at Boston Athenaeum, Burndy Library, Columbia, Harvard and New York Public Libraries.*

***Ammannati had already designed that gem of bridge architecture, the Ponte Santa Trinità across the Arno at Florence in the 1570s, and the Pitti Palace, also at Florence. The bridge was blown up by the Germans in August, 1944, and now is being reconstructed.*

shifting the prone obelisk to its new location.

The difficulties that Fontana faced were the great weight, the great bulk and the fear of breaking this fragile stone in moving it. It weighed half again more than the obelisk now in New York. Tho simple in concept this scheme could come only from an experienced hand and its success or failure depended on translating the project from its planned stage to the many details necessary for its execution.

First, the edict of authority was issued so that Fontana could obtain the required men, draft-animals, timber, equipment, subsistence and rights-of-way, and at the same time be free from litigation due to possible damage incurred in the operation. The following is the document as issued by the Pope and as based on his sovereign right of eminent domain.

"We, Sixtus V, grant power and full authority to Domenico Fontana, architect of the Holy Apostolic Palace, in order that he may more easily and more quickly transport the Vatican obelisk to the Piazza of St. Peter, to make use, as long as this removal lasts, of whatever workmen and laborers, with the apparatus that may be necessary, of whatever kind it may be, and when in need to compel anyone to lend material to him, or sell it, he, however, satisfying them with due compensation.

"That he can make use of all the boards, timbers, and wood of any size, which are in places convenient for his needs, regardless of to whom they belong, paying, however, the due price to the owner of this lumber, in accordance with the decision of two arbitrators chosen by the parties; and that he can cut, or have cut, all the wood which may in any way belong to the church of St. Peter, its chapels and canons, particularly in the ground of the Campo Morto, or of the Hospital of San Spirito in Sassia, or of the Apostolic Chamber, without making any payment, and he can carry this wood to whatever place he desires; and let out to pasture the animals used in this work without incurring any punishment, making up, however, for damage done, according to the decision of experts chosen for this purpose.

"That he can buy and carry away the abovementioned articles and anything else necessary from any person whatever, without paying excise tax or customs duty of any kind.

"That he can, without a license, or permit, get together in Rome or in other cities and neighboring places any amount of victuals for his own use and that of his workers and animals.

"That he can requisition and carry away from wherever he finds them, capstans, ropes and cords, whether loose or fixed, undertaking, however, to repair them and make them whole, paying a due recompense; and that in the same way he can make use of all the instruments and apparatus belonging to the edifice of St. Peter, and can order the agents, representatives and officials of the said building in a due space of time to make free and clear the Piazza around the obelisk so that it can be removed and to accommodate him in whatever way necessary in this undertaking.

"That he can (if it be necessary) tear down, or have torn down, the houses near the said obelisk, deciding first on the way in which to compensate the owners for the damage.

"Finally, authority is given to the said Domenico Fontana to do, command, execute and carry out any other thing necessary to this task, and, moreover, that he, together with his agents, workers and domestic servants in any place and at any time may carry any force of arms necessary, except those prohibited, all the magistrates and officials of the entire es-

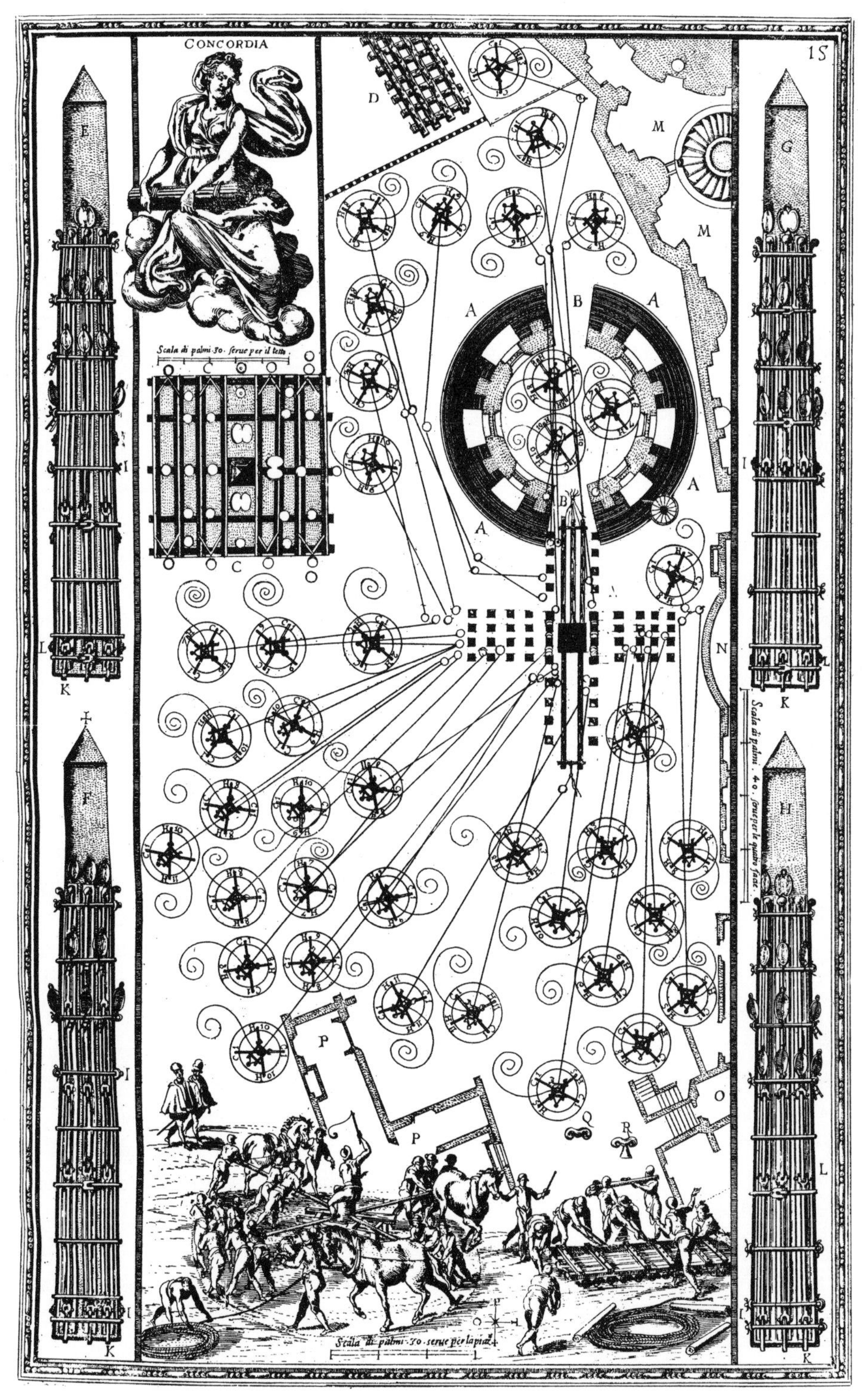

FROM FONTANA, TRASPORTATIONE DELL' OBELISCO, 1590

A plan view of the operation shows the obelisk section; the location of the tower timber-footings and the position of the capstans, including the three within the circular sacristy. The four side views show the disposition of the blocks in which E is the north face, F the east, G the west and H the south. D, at the top, is the carriage ready to be slipped under the lifted shaft.

tate of the church are commanded to aid and help the said Domenico Fontana in the abovementioned things, as are all others subject in any way at all to the authority of the Holy See, of whatever rank and condition, under pain of Our displeasure, and a fine of 500 ducats to the Treasury, and other punishments at Our discretion. No one shall dare impede, nor in any way molest the carrying out of this work of the said Domenico, his agents, or workers, but on the contrary without delay or any pretext, everyone shall help, obey and support him; anything to the contrary notwithstanding. Given at Rome in St. Mark's this 5th day of October, 1585."

In less than a month from the date of the appointment of the commission, the hundreds of plans had been examined, the winning plan selected, authority granted, and on the 25th of September 1585, Fontana with 50 men moved upon the new obelisk site, chosen by Ammannati and della Porta, and set off the working area. A pit for the foundation 45 feet square and 25 feet deep was dug in the Piazza di San Pietro in the center of what later became the famous circular colonnade. He found the soil soft and wet, and thereupon, using pile drivers, drove oak piles 9 inches in diameter and 20 feet long and had them capped at the top with a layer of peeled chestnut beams which Fontana knew would not rot in moist soil. A first course of concrete consisting of an aggregate of finely crushed basalt, flint and stone, broken brick and cement was laid down, with a sound mortar made of lime and clay which set very hard. As an appropriate observance votive bronze medals honoring the patron Pope in addition to two travertine caskets, containing twelve medals each, were laid in the concrete.

Fontana's primary engineering element was a twin timber tower erected on either side of the obelisk as shown in A on page 19. This tower, his *castello*, was designed to support the pulleys and tackle which were gently to lower the obelisk into its prone position preparatory to rolling it into its new location, then to use it again to raise the obelisk over its re-located pedestal. The tower consisted of two sections of four main inner columns each rising vertically about the obelisk to a height of about 92 feet. Each of these columns, in turn, was composed of four timbers 20″ x 20″ square so that the total section per column was 40″ x 40″, or 11.1 square feet. The timbers came from Campo Morto 20 miles from Rome and were dragged in by teams of oxen. These timbers were assembled in lap joint and were held together by 1½″ diameter iron bolts spaced at nine foot intervals and also by a series of iron bands placed between the bolts and made tight by driven wedges. These bands did double duty of holding the planking together and providing a hitch whereon to fasten the hoisting tackle. To compensate for shrinkage of the timber during the operation the bolts were made with slots in their heads to accommodate wedges which were driven tight as the operation proceeded. Alternate with the iron bands there were also placed bands of rope lashed at the points where the timbers were bolted and these were also kept tight with wedges. This type of assembly made possible the erection of the tower for lowering the obelisk, then for disassembly and reassembly of the tower for the raising phase at the new location.

The main vertical columns on either side of the line of travel of the obelisk were braced by a group of 16 inclined shores set in rows of four (thereby forming four parallel bents) each set of shores slanted inward at the same angle, thus supporting each of the main columns at four points along its length. The tallest shores were more than 94 feet long and were joined to the vertical columns to support the main yoke over the obelisk. The next lower brace met the column at a distance of two-thirds of its vertical height, while the remaining two braces were joined at one-half and one-third the column height. The base of the longest of the set of similar triangles thus formed was 25 feet 6 inches. In the engravings* used by Fontana to illustrate the towers, members are often omitted in the interest of visibility. The two outer sets of shores were spliced and had built-up sections held together by iron bands and rope lashings but the two inner shores were single, unbroken timbers. In addition, there was bolted a stiffening set of struts and diagonal sway braces. These stiffened the towers against lateral thrust. To brace against longitudinal thrust, Fontana designed a set of four shores at each corner, which were set in line in the direction of the eventual motion of the obelisk. These were set at greater distances than the lateral shores and consisted of only four inclined parallel units each. The plan of the tower footing is shown below the circular sacristy of St. Peter's on page 27. The top of the tower was crowned by a yoke consisting of four king-post struts designed to hold the 18 single and 3 double-blocks which were to carry the complete weight of the obelisk during the lowering phase. The disposition of these blocks is shown in the upper left hand plan on page 27. To aid in the lifting of the monolith prior to its inclination and lowering onto its carriage set on rollers, five great levers, each 51 feet long, were pivoted in the freeway, three such levers were combined and pointed westerly through a large aperture cut into the circular sacristy wall. The other two combined levers pointed easterly in the direction of the apex of the obelisk when brought down. The disposition of levers and capstans is also shown on this engraving.

In his book, Fontana explains how he measured the obelisk which he found to be 83 feet high and as having a base of 9 feet 2 inches square and 5 feet 11 inches at its upper section, terminating in the pyramidion 4 feet 3 inches high. He then considered the idea of dividing its form into several geometric elements. The volume of the parallelepiped, formed by considering the top area as a base and the obelisk height as its length, was then calculated. To this was added the volume of the quadrangular pyramid composed of the four corners, having the same height as the obelisk and a base formed by the difference between the top and bottom lengths. He then added the volume represented by the four remaining wedge-shaped sections, and finally the volume of the pyramidion. Comparing this mass to the weight of a stone of approximately the same material, he computed the obelisk weight to be 681,222 pounds. He then assumed that a capstan powered by four horses could lift about 14,000 pounds, and he decided that 40 such capstans could raise over 80% of the obelisk weight. He therefore depended on the five timber levers to supply the additional lifting power in raising the shaft from its base.

As can be seen from the diagram on page 27 these 40 capstans were arranged about the obelisk and yet room for three had to be found within the circular sacristy to the west of the tower. The capstans were then connected by stout hemp rope made from Foligno

**These beautiful engrdvings were the work of Natale Bonifazio, born at Sibenico in Dalmatia 1538, died 1592.*

fibers almost three inches in diameter and about 750 feet long. Extended runs of rope were made to pass over rollers set upon the ground to keep them from being fouled. Fontana personally concerned himself with the rope fabrication knowing the importance of this item. Eight heavy ropes, 1500 feet long, guyed the tower at its top. In determining the diameter and tensile strength of the rope, Fontana again made his calculations from the premise that the steady pulling force of a horse would average 100 pounds. Four such horses at a capstan having a 9-to-1 ratio, gave a pull of 3600 pounds per capstan and when used on 2-to-1 ratio blocks gave a pull of 14,400 pounds. Since Fontana tried and could not break the rope selected with a pull of 50,000 pounds, he concluded that the margin of safety would cover not only variations in the rope manufacture but also the peaks of tension imposed on any one capstan when other capstans should run slack. The even burden to be put on the capstans was made certain by requiring the capstans to turn only three times and then to pause for hand-testing the proper tension on each rope. At each capstan Fontana noted the number of men (H for *homini*) and the number of horses (C for *cavalli*) intended to operate the mechanism.

In advance of any operation there was a thoro planning and rehearsing. The arrangement of pulley blocks was studied to correspond to the plan and to the location of the capstans, both block and capstans being given corresponding numbers. The various pulley blocks were designed and made as shown on the top of engravings on page 22. The largest were of wrought iron 5 feet 2 inches long with six metal sheaves set in two layers of three each, the upper set of sheaves being larger in diameter so that when reeved, the upper rope would clear those beneath them. There were double-sheaved pulley-blocks having wooden and iron sheaves set in their sides. They were, however, always armored at the sides with iron bands and had iron rings. Fontana also used single and double pulley-blocks. In the same view are shown the iron bands, bolts and wedges for binding the timbers to form the main tower columns and shores.

The obelisks then in Rome had been subject to so much abuse by the movers and the vandals that most had suffered breaks and cracks. Fontana was therefore particularly careful in anticipating the development of cracks in his obelisk by avoiding the strains that would have expanded such cracks into breaks. He thereupon ordered that the shaft be protected by reed mats to cover the delicate surface and then encased these by an overall planking two inches thick. To grip the shaft and to lift it, Fontana placed rectangular iron bars under the shaft. These were U-shaped, acted like saddles under the shaft and had bar extensions along the four faces. These bars were two inches thick and 4.4 inches wide. There were four on each face (altho frequently shown in the engravings as three). These 16 jointed vertical bars were

(Plates I and II, right): Two engravings prepared during the moving operations of the obelisk. In the first, at the left, is shown the obelisk partially buried and with the gilt ball at its top, as the Romans left it. At the right (less the center embellishment) is the shaft as Fontana re-erected it. The timber tower, iron lifting-bars, tackle and carriage are shown, center. The second plate shows the piazza during the initial lifting operation in April, 1586. In addition to the unfinished St. Peter's, left, there is seen a view within the circular sacristy and, adjoining it, Fontana's tower, surrounded by the capstans and crews. These plates were engraved by Natale Bonifazio (born in Sibenico in Dalmatia in 1538, died in Rome, ca. 1592) and Giovanni Guerra (born in Modena in 1549, died in Rome in 1618). They were printed by Bartholomeus Grassius for general sale during the great event. Reproduced by courtesy of The British Museum.

MDLXXXVI.
Romæ
Gug.a quando cala à basso, è posti dall'altra banda lo tirrarano inã
zi mentre s'alzerà:
F. Strascino sopra li Curli, sul quale si colca, & ha da essere strascina
ta la Guglia:
G. Letto di trauamenti collegati insieme, sopra il quale lauorarano i cur
li per tutto il camino mentre la Gug.a sarà strascinata per forza d'al
tri argani non posti nel presente disegno, quali sarano piantati
in luoghi accomodati al proposito del passaggio uerso doue ten:
dono. i. canapi segnati gg. attaccati alla Gug.a colcata:
H. Gug.a tirata per condurla nel luogo, doue ha da esser drizzata.
I. Inbragat.a della Gug.a dal piede alla Cima di ferri grossi, quali ha
uerano gli occhi, & cerchi a guisa di maschietti: ma p.a sarà fasciata
di store doppie: acciò non resti segnata.
L. Castello aperto, che mostra la parte interiore con la Gug.a attaccata
alle Taglie.
M. Traui, che fanno lieue, sarano 12 per aiutare ad alzare la Gug.a
N. Gug.a posta in opera sopra il suo piedestallo. è zoccoli, chi erano
sotterrati, eleuata dal mattonato della piazza sino alla somita
P. 155. m. 37. con gli scalini, che uanno atorno.
Scala ouer misura della sudetta Guglia fatta con piu nedute in 3 mis.re
Palmi — 5. I. 10 15 20 25
Palmi 5. II. 10 15 20 25 30
Palmi — 5. III. 10 15 20 25 30
A Lettori
Spero, benigniss. lettori, con la gra. di Dio, donarui un uolume,
doue diffusam.te si dichlerarano tutte l'operationi, con le Ma:
chine et instrumenti à parte, a parte in disegno, & tutte le circon:
stantie loro à giouamento, e sodisfattione uniuersale, dopo, che
l'opera sarà peruenuta al desiato porto.
DOMENICO FONTANA DA
MILI DIOCESI DI COMO
INVENTORE ET CONDVTTORE
DIVO CAES. DIVI
IVLII. F. AVGVSTO
TI. CAES. DIVI AVG
F. AVGVS. SACRV
LIII

PLATE I

FABRICA NVOVA DE LA CHIESA DI S. PIETRO
DOMENICO FOTANA DA MILI DIOCESI DI COMO ARCHIT. E CONDVTTORE.
In Roma. Natal Bonifacio de Sebenicco Fece. d'Agosto. 1586.
DOMINICI FONT. EFIGIES
AN. AGEN. XXXXIII
Cum Priuilegio
Misura del Palmo che si usa in Roma

PLATE II

girdled by nine horizontal iron bands joined on opposite faces by knuckles and pins similar to the modern eye-bar; the vertical U-bars terminated with an eye at each end. To these eyes were connected the extensions of the vertical bars and these were sectionalized into two or more lengths because their length above the lowest eye would rise upward about 50 feet, making a single bar too heavy to handle. At the very top these bars were terminated in bulbous outward-pointing tips. The combined weight of the protective planking, the blocks, the ropes and bars rose to 56,459 pounds, which were added to the weight of the obelisk, making the total 737,690 pounds which had to be lifted by the ropes and levers. On opposite sides of the tower the main timbers had scantlings attached to them forming ladders that reached up to the top.

Fontana knew that the success of the enterprise depended on the uniform distribution of the great weight over the many ropes holding the obelisk so that no rope would be strained more than any other. Those who had competed in winning the commission and had lost to Fontana now jeered at his plans and claimed that undoubtedly the ropes would fail because of unequal stresses. Fontana thereupon placed a trumpeter near the engineer's observation station and the entire crew was drilled in its task. A trumpet blast was the signal for the capstans to begin turning and the sound of a bell was the signal to stop. Thus, no confusion could result. In addition to the normal crew, each member of which had a specific duty, provisions were also made for the unexpected emergencies. Eight or ten of the most experienced workmen were distributed about the working area to act as general foremen. Two foremen were assigned to each capstan and 20 men were assigned to emergency calls; their task was to quickly bring rope, blocks or replacement parts when failure threatened. These men were stationed at the door of the main stores building so that no capstan-tender would have to leave his post. Twenty horses and drivers were also placed in reserve, ready to move toward any point of need.

When the time came for inclining the obelisk after raising it above its base, it was planned to draw the shaft at its foot towards the west into the opening pierced through the sacristy wall. The obelisk foot was to be rested on a timber carriage laid on rollers. Fifty-three men were assigned to the levers in the raising process, 35 to the westerly levers and 18 to the two levers pointing east. Some were to pull directly on the ropes attached to the levers, others to bear on the capstans attached by rope to the levers. Additional men with sledges and mauls clambered thru the tower and obelisk driving wedges to tighten the binding members and stiffen the system. As in any similar modern job, these men wore metal helmets to protect them from objects falling from above.

"Acque alle Funi!"

For further smoothness in the procedure of the work without interruptions, two additional precautions were taken. It was arranged to distribute baskets of food to the scattered teams so that no one would have to leave his post. In addition to the tall barricade enclosing the working area, stringent orders were published that prohibited anyone beyond the barricade from passing the barriers or hindering the workmen in any way, nor might they speak, spit or make any loud noise, under penalty of severe punishment, "including death." This was intended to make it easier for orders to be heard over the noise of 900 moving men and 74 horses and over the scrape and rumble of ropes, blocks, capstans and timber. Police were assigned to carry out this strict order and the resulting silence gave rise to one of the prettiest legends in engineering history. This story has, thru the last two centuries, become so much a part of the general story of the erection of the Vatican obelisk that, in permitting its refutation, one owes to tradition, at least, the kindness of its repetition.

It seems (so the legend goes) that at one point of the operations the silenced multitude noted that all was not going well with the strained ropes. One version has it that the speed and strain of winding caused the ropes and blocks to get dangerously hot, another version has it that they stretched so much that they piled up on the capstan drums, the leverage ratio dropping so low that the horses and men could no longer continue to pull. At this critical moment, a lusty voice rang out from someone in the silent crowd—it was the voice of an old Genoese sailor named Bresca di Bordighera, and it called out "*Acque alle Funi!*" —Water on the ropes! Thereupon water was poured on the failing ropes, they grew taut, they held, and the obelisk was saved! Instead of condemning the spirited and brave sailor to death the legend has it that honors were bestowed on him and his family, and that to this day his descendants have the exclusive privilege of selling palms used in the procession on Palm Sunday in St. Peter's.*

**We are indebted to Dr. Cesare Olschki of Florence, bibliophile and authority on the Renaissance, for locating and translating a monograph by Prof. Christian Huelsen treating this subject* (ROMA, Rivista di studi e di vista Romana, *Anno I, pages 412-418). In this it is pointed out that neither Fontana nor any contemporary notice referred to what would be a most notable occurrence. The absurdity of securing, distributing or applying sufficient quantities of water with the primitive hydraulic equipment then known, is disregarded. The earliest reference to water on the ropes of the Vatican obelisk was traced by Huelsen to Volkmann writing on Italy in 1750. The same story with different actors and related to the raising of the obelisk at Atmeidan, one of the two carried to Constantinople, was described in manuscript in 1555 and published in 1581, five years before the raising of the Vatican obelisk by Fontana. This was a report made by Ogerio Chisleno di Busbeke, ambassador from the Viennese court of Emperor Ferdinand I to the Sultan at Constantinople, a gentleman known for his scholarship and archaeological interests. In the printed version of his book* De Legatione Turcica, *Amsterdam, 1581, Busbeke relates, "This story is also confirmed in a similar report by the traveler Rinaldo Lubenau who visited Constantinople in 1587-8 and writing to his native Koenigsberg, repeated the main elements of the account". This was published in the city's archives in 1912-20. In all probability some chronicler transferred the Constantinople version of the obelisk-raising story to the more familiar Vatican setting and thereby gave us today's much circulated story. Similar stories had been told of the moving of the colossal stone forming the pedestal of the equestrian statue of Peter the Great in St. Petersburg weighing 600 tons.*

The Obelisk is Lowered

THE DATE for the lowering operations of the obelisk was set by Fontana as April 30, 1586. A last preparatory task was the removal of the ancient metal ball still at the obelisk top, placed there by the old Romans and believed to contain the ashes of Julius Caesar. The ball was removed but no ashes were found there. The tower was complete, blocks were in position, capstans and ground tackle fixed in their places. The men were divided into their respective crews, assigned to their posts and again drilled each in his special task. Their equipment and supplies were distributed and last minute checks were made. An air of expectancy fell upon the metropolis.

The day prior to the scheduled operation was given to devotion and prayer. The workmen attended confession and were given the sacrament. Two hours before dawn of the 30th, two final masses were celebrated and by sunrise every man and horse was at his post ready to proceed. The morning was a typically clear and serene Roman spring day and all Rome moved towards the Piazza to witness the event that had held their interest so deeply and for so long a time. The great personages of Rome and the nobility of all Italy were present. The majority of the College of Cardinals and the city officials were there. Upon roofs and windows facing the square and upon the walls and scaffolding of the incomplete St. Peter's there were crowds of onlookers. The city police was reinforced by the Swiss Guards and a detachment of cavalry kept the swarming populace in order.

Fontana's first command was for the workmen to kneel in prayer with him, and a Pater Noster and Ave Maria were said. He then rose and upon his signal the trumpet sounded and over 900 men and 74 horses bore on their lever-arms. The gear groaned, ropes tightened and the great stone budged with a rumble that Fontana describes as like an earthquake. The bell was then rung and positions were held as the foremen inspected the equipment. It was found that the top one of the horizontal bands had burst. Evidently the radial component of tension on this band had been underestimated in arrangement of the blocks. Repairs were made and in twelve additional moves the shaft was raised more than two feet above its pedestal. This permitted the timber carriage, moving over rollers, to be slipped into the gap, completing the first phase of the operation. By 10 o'clock that evening the quitting-signal was sounded. The success of this first step was celebrated by the firing of a signal gun which was answered by a burst of artillery from the city's batteries and great joy was shown by the multitude.

Upon re-examination of the results of the first move in the operation it was found that not only the top band had broken, but that all nine bands had suffered somewhat—some had slid from their positions, others had twisted or had developed breaks in their sections. To prevent further slippage, rope slings were passed under the obelisk foot, for it was found that none of the rope lashings had suffered as had the iron ones, and this added precaution had saved the old stone. It was also found that some of the vertical bars had also been broken at their eyes. Evidently the bars had not been adjusted to share the load equally as no adjustment mechanism (turnbuckles or wedges) had been provided and the "cascading" of such breaks had not yet been a province of study. The result was the shearing of pins and jaw-members in the hand-forged, and therefore ill-fitting, elements. Fontana did not understand "shear" and was mystified by the clean break that looked "as if cut by a knife".

The main shaft had rested upon four bronze corner feet or astragals* and it was the next task to remove these while the great stone remained steadily suspended on the ropes. These solid bronze castings weighed about 600 pounds each and their removal required four days and nights. Evidently the old Roman engineers had fitted the shanks projecting from them into holes in the base and these could not readily be pried loose and finally had to be cut away. From an examination of these feet, Fontana concluded that the Roman engineers had had some serious problems in setting the obelisk. Since the feet on one side were flattened, he presumed that that was the side onto which the obelisk had been first lowered from its ramp position and then pivoted upward in a vertical arc until it was upright, finally resting on all four feet. Fontana had found the obelisk not in a truly vertical position, it having tilted seventeen inches to the north, a position that cor-

*Q and R, lower right page 27, show the bronze Roman astragals. Fontana states that Q weighed 450 lbs. and that R, cast in one piece with its shank, weighed 600 lbs.

rected itself when he first swung the shaft clear from its pedestal. This concentration of stresses in the lower part of the obelisk when first placed on its heels by the Romans may well have split off a portion of the shaft. That this may actually have happened is gleaned from an account of such a split given by Pliny in the first century A.D. which stated that the obelisk was broken during the process of erection. Fontana, carefully examining the shaft and comparing measurements with equivalent measurements and ratios of other obelisks in Rome, pointed out that the ratio of the height of the pyramidal top to its base is in fact only half of the ratio of that of the others. Further, the finished tool marks of this pyramid are different and rougher than those on the side, indicating a probably later modification. A third indication of change in the shaft was that its overall height ratio to base dimension is less than that of the other obelisks, again indicating that changes had been made on the stone since the Egyptians had first fashioned it.

In the plan layout of operations on page 27, there is shown (top center, D) the carriage on which the prone obelisk was to be lowered and on which it was to rest during its transportation from the old to the new location. This carriage was made of four timbers almost sixty feet long, cross-braced by timbers six feet six inches long and framed into the longitudinal timbers.

For the lowering operation three new steps were taken. The first of these was to shift the system of pulleys from the east face of the obelisk to the other three faces so that this face, being the bottom one when resting on the carriage, would have no encumbrances in the descent. The second step was to introduce a sliding brace, having its top bear upon the underside of the lowering obelisk and having its foot slide outward as the stone came slowly down. This angular brace, hinged at its top, is shown as C at the right of the obelisk on page 35. The brace consisted of four timbers 44 feet long having their upper and lower ends terminating in iron-shod rounded ends. The upper ends pressed against the middle of the obelisk and were held together by a 4.4 inch iron rod that was held in place by an iron stirrup banding the obelisk. In this way the brace could pivot but could not otherwise move. The lower end of the brace had sheathed, rounded ends so that these ends could slide outwards along the tops of prepared timbers. This outward sliding was controlled by having anchor ropes and tackle attached to an iron bar passing thru the four lower brace ends, and as the obelisk was lowered and the braced footing moved outwardly, the motion was checked by the slackening of the anchor ropes. In this way a stable triangulation was introduced that permitted gradual and constant adjustment of the lowering rope system; in fact, this moving-rake shoring made the descent almost independent of the rigging. However, its effectiveness was reduced as the obelisk became more horizontal so that this brace was replaced by a shorter brace, and finally, with the obelisk quite prone, by a series of seven piled-up blocks and wedges to conform to the taper of the obelisk's lower side.

The third innovation was a group of four blocks that were attached to the foot of the obelisk and the ropes from these were passed thru a corresponding set of blocks positioned on the west face of the sacristy. The purpose of this was to be sure that the base of the obelisk would be properly drawn thru the opening made in the eastern sacristy wall as the obelisk moved away from its vertical into its horizontal position. This would also help keep the main ropes in a vertical position thereby avoiding lateral strain. However, these elaborate precau-

FROM FONTANA, TRASPORTATIONE DELL' OBELISCO 1590

The foot of the obelisk rests on the carriage and is moved on rollers to the left. To the middle of the great stone there has been added a raking shore whose foot also rests on the carriage but moves outward towards the right. The iron eye-bars on the faces of the obelisk are held by surrounding bands to which the tackle is fastened. For greater clarity, not all of the main tower shores are here shown.

tions seemed quite unnecessary because when the shaft was half-way down, there was shown a tendency for the carriage on its rollers to slide westward without any aid; in fact, the slide had to be controlled by rigging up ropes in the opposite direction from that originally intended. To avoid the danger of impact developing from

FROM FONTANA, TRASPORTATIONE DELL' OBELISCO 1590

The obelisk has been lowered and rests on the carriage ready to move eastward (see rose-of-the-winds, top left). To the yoke at the top of the tower are fastened the main blocks. In the foreground lie the rollers on which the carriage will be moved.

a jerky or too rapid motion of such a great and fragile weight, it was deemed necessary to add a bridle by attaching five additional blocks to the top of the obelisk and to connect these by rope to five equivalent pulleys suspended from the top of the sacristy.

It required eight days to make these preparations for lowering the shaft, thereby bringing the date of lowering up to May 7. In this, as in the initial raising of the shaft, the same signals were used, but in reverse order. At the trumpet blast, the main capstans paid out rope while the capstans connected at the base wound theirs up, thereby pulling the base westward. As the bell sounded work stopped, the gear was examined, and the ropes were tightened equally. The operation ran smoothly with no unexpected occurrences. By 10 o'clock that night the obelisk had been completely lowered and rested properly blocked up on its carriage. The master engineer of this operation then received his due reward by being accompanied to his home by an honor guard of drums and trumpets.

Moving and Raising the Obelisk

THE NEXT four days were required to prepare for the next phase of the operation, transporting the obelisk. The capstans, rigging and tower had first to be dismantled and this required rolling the sleeping monument out of range of possibly falling iron or timber. With bolts loosened, the tower members and rigging were then removed to the new site and work was begun clearing the obelisk pedestal of its 1500 years of accumulated earth and debris which had buried it almost up to the base of the shaft. The solidly-packed earth about the stone pedestal made it possible for Fontana to lay the planks for the rollers and carriage right under the obelisk itself.

Fontana planned to replace the obelisk pedestal in its new position in a condition as much like the original as he could. He therefore moved the metal corner astragals and then the pedestal stone, which stood eight feet six inches high and was nine feet six inches square and weighed 55½ tons. Beneath it and occupying a narrower space was a rougher stone weighing 63 tons. The next courses were of white marble connected by iron clamps encased in lead. Fontana was surprised to find the iron of the clamps in a perfect state of preservation. The lowest courses were of travertine forming three steps that rested on decomposed concrete. Because the top stone was damaged in removing the astragals, two inches of the pedestal had to be cut away. In the new pedestal, between the two lower strata of marble, was placed a new stone bearing the names of Sixtus V and Fontana and giving an account of the transportation. On the face of the pedestal's lowest tier was cut *Dominicus Fontana ex pago Mili agri Novocomensis transtulit et erexit*, but this inscription, last recorded as having been read in 1830, has since been worn into complete disappearance altho it does appear in the engravings in Fontana's book and in others. Here is again seen how the printed word can be more enduring than stone.

The failure to increase the courses beneath the main pedestal stone led Fontana to conclude that each stone had formed part of some foundation that the old Roman engineers did not disturb when they first set up the obelisk in 41 A.D. The pedestal stones in their original relationship were re-laid in their new positions above the prepared tile foundation, making the total pedestal height 27 feet.

A contemporary fresco painted on the walls of the Vatican Library shows Pope Sixtus V during his coronation on the Piazza of St. Peter's prior to the time that the old buildings facing the Piazza had been torn down to give place to the grand facade as we know it today. This fresco also shows the obelisk in its old position behind the sacristy. The distance from the old position to its new in the center of the recreated Piazza is only about 275 yards and covers an elevational drop of nearly thirty feet. This was of great advantage to Fontana for it enabled him to bridge the distance by a causeway of increasing height so that the obelisk could be swung right over its newly reconstructed pedes-

FROM FONTANA, TRASPORTATIONE DELL' OBELISCO, 1590

The obelisk on its carriage is moved over the earth-filled causeway. The iron bars on the obelisk faces and their binding bands are here clearly shown. The direction of motion towards the new site is easterly over a 275 yard length.

tal without having to be raised the corresponding height. This causeway, as shown above consisted of earth fill, retained by a shored tim-

DISPOSITIONE E VEDUTA GENERALE DELLE MACHINE CHE SERVIRONO PER A

FROM ZABAGLIA, CASTELLI E PONTI, 1743

The scaffolding used to restore the Vatican obelisk in 1732 when the bronze embellishments were added.

FROM CARLO FONTANA, TEMPLUM VATICANUM, 1694

The view of the Piazza of St. Peter's showing the section and terminus of the causeway looking eastward. Upon the broadened section of the causeway Fontana has re-erected his tower and from its top is hung the tackle that will raise the obelisk. Each of the forty-eight capstans is in position and the ropes are taut. In the center position, in the command tower E, Fontana issues his commands. By nightfall the ancient stone will hang vertically over its reconstructed pedestal.

ber crib. For its main length it was 37 feet wide at its top and 73 feet wide at its base. At the terminus it was 27 feet high and 92 feet wide at its base. About the new pedestal the causeway swelled out to a top width of 70 feet so as to provide room for positioning the tower that was to swing the obelisk back into its vertical position.

The preparations for resetting the obelisk over its new base consisted of reversing the procedure in having it lowered. The tower timbers were dragged to the new site and re-erected on the broadened section of the causeway. On the tower were hung pulleys and riggings for the raising operation and the rope and pulleys were attached to the bands of the three free faces of the obelisk after it had been moved on its carriage and rollers to its position over the pedestal. Because of the spread length of the obelisk in its prone position and the increased stresses in the raising (as against the lowering) process, the number of horses was increased from 74 to 140. Of the 48 capstans set out for the raising process, 40 were assigned to the lifting tackle, and in the east end four were intended to drag the obelisk base forward to the central position. The four on the south portion of the field were not used.

On September 10, 1586 dawn found every detail ready. The crew of 800 men was again confessed and under the same system of trumpets and bells, the capstans turned between pauses for inspection and adjustment. At the east end of the ramp a command tower was built for Fontana and his deputies. With the experience gained in the lowering phase the raising moved much more quickly and smoothly. When the obelisk was at about 45° angle, it was buttressed by timbers and the tackle was made fast. The crews then were permitted to eat and rest. At sundown that day the shaft was hanging vertically over its pedestal after 52 pulls and pauses. The joy of the great crowd and the holiday atmosphere of the occasion gave rise to spontaneous jubilation.

The next day, again as in the lowering stage, levers were in use to raise the shaft over the pedestal. The carriage had been made narrower than the obelisk base so that as it hung over the pedestal there were clear vertical spaces under the north and south faces. With the obelisk raised, blocks and wedges were driven between its base and pedestal and the carriage was thus free to be rolled out of the way. The capstans were then tightened and the bronze astragals now in the form of gilded bronze lions*, the work of the sculptor Prospero Bresciano, were set in stress-equalizing lead at the four corners. Plumb-bombs showed that these astragals were not level, thereby requiring metal shims to be inserted until the shaft stood perfectly vertical. In eighteen days the tower, cribs and scaffolding had been cleared and on September 28, 1586 the obelisk appeared in full view and was duly consecrated. It had taken Fontana just one year from the time of winning his commission to complete his task and to leave the obelisk standing practically as we see it today. Only the bronze eagles at its base and a more elaborate balustrade were later added.

The stone was then purged of its pagan associations by a consecration of the cross in a procession ordered by the Pope. He ordered a meeting of prelates to consider means of erecting a crucifix upon the obelisk in order to glorify the cross by making the obelisk its pedestal. This consecration also was to remove from this

These lions, symbol of the pontifical family Peretti, represent an architectural curiosity. In order to show two lion bodies on each obelisk face it became necessary to model the lion at each corner with one head and chest but with a body extending in both rectangular directions. Thus the four corners of the shaft rest on lions having four heads but eight bodies.

ancient stone the pagan superstitions of its past. Between obelisk and cross there were introduced the pontifical devices of the five stacked peaks and above them the many-pointed star.

Rewards and honors were then showered upon Fontana. He was made a Palatine Count and a Knight of the Golden Spur and he was also granted a pension of 2000 gold scudi as well as the recipient of an immediate gift of 5000 more. He received, gratis, all the wood and equipment left over from the operation. Architectural commissions also were thrust upon him. He was required to supervise the restoration and erection of three additional obelisks in Rome. One, a few feet shorter than the Vatican obelisk, was re-erected by him in the Piazza del Popolo, the second he placed in the Piazza di San Giovanni Laterano. This is the tallest obelisk known, even tho we think that part of the base of the shaft had, at some time, already been cut away. The third is behind the church of Santa Maria Maggiore and stands 48½ feet tall. However, the pedestal upon which it stands is 17 feet high, thereby giving it the appearance of good overall height. Two of the above, those at San Giovanni and Santa Maria had been buried and lost, but once in a conversation between Sixtus, Fontana and Mercati* it was urged that more obelisks be hunted. Fontana, using an iron bar, probed the fill in the Circus Maximus and came upon two obelisks, both fallen and both broken into three parts. Fontana joined the broken sections with the use of stone dove-tail mortises.

In addition, Fontana was assigned the task of designing the Vatican Library and the Papal palace on the Monte Cavallo. He also completed the architectural masterpiece of Michelangelo—the dome of St. Peter's.

He engineered the conduction of the water supply Acqua Felice and erected several fountains, including one of the pair near the Vatican obelisk. In the introduction to his book he lists 35 major architectural and engineering accomplishments, any one of which would make an accomplished architect proud. As the times required, the blending of architecture, engineering and the fine arts came in almost equal proportions, for Fontana was also charged with the setting of the bronze statues of St. Peter and St. Paul and the renovation and resetting of the ancient columns of Trajan and Antonino as well as the transportation and restoration of the classic horses by Praxiteles and Phidias.

With the death of his pontifical patron Sixtus V in August 1590, Fontana lost favor with the new Pope and became Senior Engineer to the King of Naples. There Fontana built a diversion canal, constructed a coastal road and designed a new royal palace. His final project was for an improved port of Naples, a plan executed after his death in Naples in 1607. Altho he lived thru the reign of five Popes, only Clement VIII (1592-1608) gave Fontana an architectural commission, a bridge over the Tiber at Borghetto.

The magnificent setting of the obelisk as we know it today was undreamed of by either Fontana or Sixtus V, for in 1586 St. Peter's was far from completed. When Michelangelo died in 1564 he left his grand plans for the church for others to finish. Thus, Vignola, Ligorio and della Porta, assisted by Fontana, then carried on. It was the latter two who finally completed the immense dome in 1590. At the time of Fontana's obelisk operations the facade facing the Piazza consisted of an assembly of five old buildings. These were torn down and the grand facade of St. Peter's, as we know it today, along with the additional portico, were built by Fontana's nephew, Carlo Maderno, but this was

**See Bibliography page 61.*

A contemporary fresco on the walls of the Vatican Library shows the obelisk in its new position and a view of the Piazza of St. Peter's as Fontana saw it. In the background are the unfinished walls of St. Peter's, and the odd buildings later replaced by the facade of Maderno. The grand colonnade encircling the obelisk was erected 70 years later.

not completed and consecrated until 1626, 19 years after the death of Domenico Fontana. The old facade was demolished in 1605 and the nave was lengthened in the direction of the obelisk. The familiar encircling colonnade moving out in a converging line and then embracing the piazza with the obelisk as its center was not built for another 70 years (1656-1657). It consists of four rows of 284 Doric columns and 88 pillars, surmounted by some 300 statues and was the masterpiece of Giovanni Bernini (1598-1680). Of the two fountains flanking the obelisk, one was set there by Sixtus V, and the other by Clement IX (1700-21). In 1723 a rose-of-the-winds was placed around the foot of the obelisk and in 1817 the astronomer Gilii traced a meridian curve around it.

To the balustrade and four small posts that Fontana added to protect the odd base of the obelisk pedestal against vehicles, eight more posts were added by Urban VIII (1623-1644). Under Clement IX (1700-1721) the original four columns were replaced by a larger ornamental marble balustrade and a quadrature of sixteen heavy granite circular columns surrounded by a circular travertine walk. To cover the unsightly holes that remained when the old Roman ornaments were removed, the architect Lodovico Sergardi added a gilt bronze garland of oak leaves surmounted by an eagle, on each of the four faces of the obelisk.

The thoroness and attention to details, the technical analysis of the problems, the planning and scheduling, and the promptness of execution of an approved plan marked Fontana as a master in his profession. The preparation of his report and seeing it thru the press so that succeeding generations of engineers could profit from his work, raises him further in the esteem of his peers. That he worked with ample margins of safety, yet without hesitation, once the most practical solution of a problem had been decided upon, sets him up as one whose methods can well be followed.

In using structural members he saw to it that they were amply proportioned yet light enough to be transportable by making them of built-up sections. Where possible he made a member do double-duty by using removable bolts and bands instead of spikes. He knew the stability of triangulation in structures and used bracing wherever he could, breaking large triangles into smaller ones by adding sway braces. In many cases he provided adjustments for shrinkage of structural parts by using bolts and wedges, and he divided loads among as many supports as he could, equalizing these loads wherever possible. In his book he gives detailed costs of labor and equipment, showing the economic basis of his engineering operations. He enumerated his failures as well as his successes, noting the bursting of the iron bands and the shear thru the eyes of the long U-bars. A generation before Galileo first began to determine the strength of materials by mathematical analysis and experimentation, Fontana already was determining these by calculation.

Modern Transportation of Obelisks

WITH THE fall of Rome, darkness descended on all the civilized world. In Europe the dark ages passed, but to Egypt no light has yet returned. However, an accidental find restored the interest of scholars in the past greatness of Egypt. This was the discovery in 1799 near Rosetta at the mouth of the Nile of the now famous Rosetta stone. It was a black basalt slab which was found by soldiers digging a trench under command of Col. André-Joseph Boussard of Napoleon's Army of the Nile and was used by Champollion* to decipher the hieroglyphics of the Corfe Castle (England) obelisk. The three sections of the stone contained priestly instructions in hieroglyphics, demotics (a simplified form of Egyptian hieratic alphabet) and Greek, each a key to the language of the other. It is now in the British Museum. This find electrified the world of scholarship for it made understandable all the markings on the mummies and monuments that had found their way into the western world. A new interest in a great civilization, long dead, was aroused.

In the past century three obelisks were moved from Egypt, all gifts of the head of the state at a time when the status of Egypt was that of a colony. The removal of each of these three stones is treated below in reverse chronological order.

The New York Obelisk

IT REQUIRED the passage of some 75 years from the time the British received their gift of the prone obelisk in Alexandria before this stood erect once again on the London Embankment. Twenty-five years passed between the time that Napoleon I had decided to bring back with him a trophy of his campaign of the Nile until such an obelisk graced Paris. It is therefore understandable why the very retiring engineer that brought the obelisk into New York in about a year should point to this performance with some pride.

Unlike the uninscribed shaft at the Vatican, the one now in New York has all its faces and its pyramidion covered with hieroglyphics and

**Jean Francois Champollion (1790-1832). was the founder of the Egyptian museum in the Louvre.*

therefore its history is well known. Three kings of Egypt have engraved their names and titles on its four faces, two of these kings among the mightiest of the hundreds that ruled Egypt. Yet, when Lieutenant-Commander Henry H. Gorringe, U.S.N., first saw it in the filth and squalor of the Alexandrian embankment, a prey to destruction by vandals and the sea, there was little association with the nobility it once knew.

This shaft was one of a pair that stood before the great temple of Tum in the sacred city of the sun, Heliopolis. The central column of inscriptions on the four faces of the monolith informs us that it had been cut and placed by Thothmes III, pharaoh in the 18th Dynasty (1461 B.C.), on the occasion of the fourth jubilee of his reign, that is, on having reigned forty years. To this inscription was added, some 150 years later, that of the great builder of Egypt, Rameses II (19th Dynasty, 1388-1322 B.C.), consisting of a column on either side of the original, forming three columns, and finally at the outer edge a very small inscription by Orsoken I (22nd Dynasty, 933 B.C.).

The pyramidion is also engraved on its four faces. The designer and architect of the two

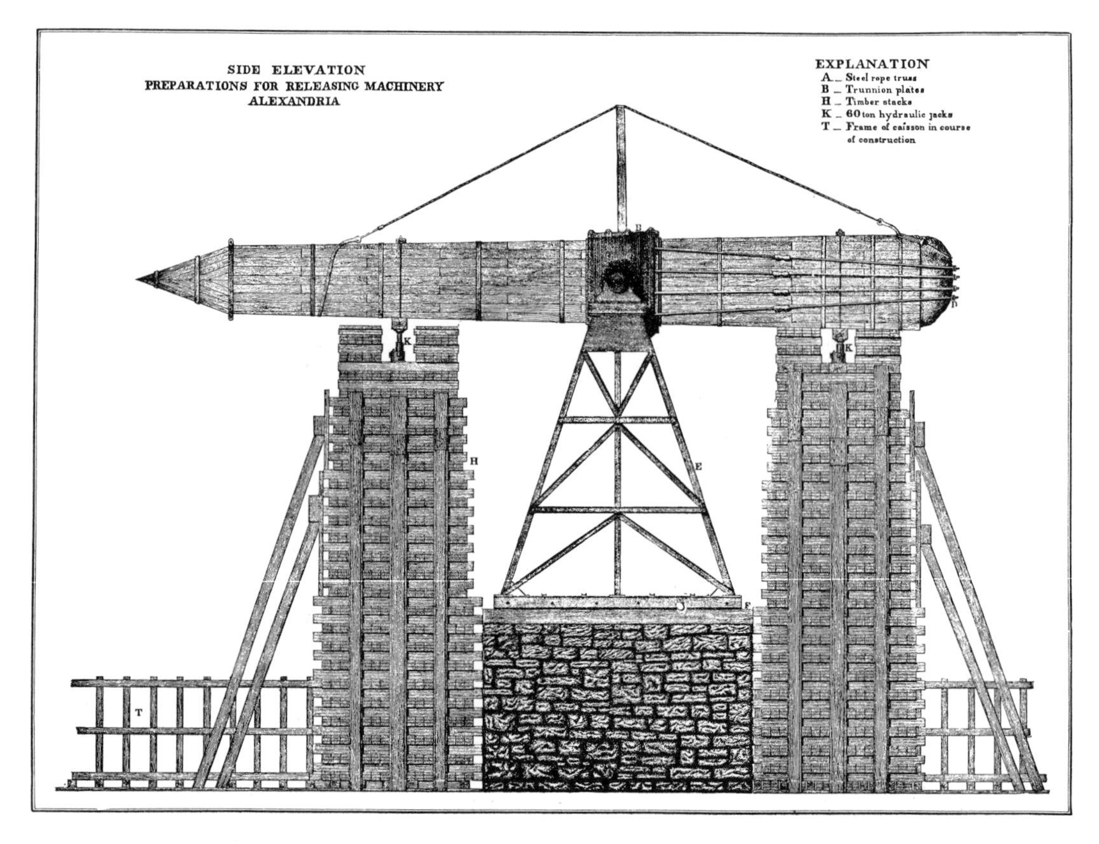

FROM GORRINGE, EGYPTIAN OBELISKS, 1882

The means used by Commander Gorringe for turning and lowering the obelisk at Alexandria. The main weight is carried by the two steel frames E which rest on masonry piers. The turning is done about the trunnions fastened to the middle of the shaft. Hydraulic jacks K then lower the horizontal stone until it can be slid into a floating caisson. The process was reversed at its New York destination.

obelisks which ultimately found their way to London and New York was Amen-men-ant*. When the Caesars became active in Egypt the pair of obelisks was moved by the Roman engineers and set up at the water entrance to the Caesareum, an arena built in the city and dedicated to the Roman Caesars, the conquerors of Egypt. There the monuments stood thru the period when the power of Rome fell and the fiber of Egypt decayed. At some time between its transportation to Alexandria and its embarkation for London, one obelisk fell to the ground leaving the other of the pair standing but leaning somewhat towards the sea. The fallen shaft was removed in 1877 and the remaining one stood alone when Commander Gorringe arrived. Commander Gorringe had been selected by the sponsors as an officer who combined the talents and experience that assured the safe transportation and erection of this heavy and fragile stone.

Even tho Mr. William H. Vanderbilt of New York had underwritten the cost of transportation to the amount of $75,000, he made it clear that payment would be made only when the shaft was successfully placed upon the chosen site. This meant that the full responsibility for the design, negotiation and execution rested on the shoulders of Commander Gorringe**.

In Egypt Gorringe found every form of impediment to the operation that the unfriendly European residents in Alexandria could put in his way. The German and French archaeologists fought the removal of the obelisk and incited interference by local authorities. They were perfectly content to let the shaft disintegrate but would not lift a finger to help in its removal. The Khedive of Egypt had presented the obelisk as a gift to America thru proper agencies, and the firm and direct action of Commander Gorringe reduced the time required to move it (estimated by the Egyptians as a century) to a little more than a year. On October 27, 1879, one month after the arrival of the turning equipment into Alexandria harbor, Commander Gorringe had the obelisk off its pedestal, thereby serving notice to the obstructionists that the stone was definitely on the

FROM HARPER'S WEEKLY, VOL. 29, 1885

Lieutenant-Commander Henry H. Gorringe, U.S.N., whose vision, technical ability and firmness made possible the transportation of the obelisk from Alexandria to New York in 1880. In this he faced difficult political as well as engineering problems.

*JULIEN, *page 99, also states that the identity of 40 of Egypt's architects and engineers is known but mentions only one other—Ra-nefer, the architect of Memphis, whose statue was dug up near Cairo.*

***Henry H. Gorringe was born in Barbados (British West Indies) in 1841. He came to the U.S. in his early youth and joined the marine merchant service. During the Civil War he joined the U. S. Navy in 1862 and served as master's mate in the severe fighting in the Upper Mississippi with Porter's Flotilla. He remained in the service after the war and was commissioned Lt.-Commander in 1868. In 1871 he commanded the sloop Portsmouth, in service with the South Atlantic Fleet. He was author of several books on sailing and navigation. After directing the transportation of the obelisk to New York, he resigned his commission in 1883. Two years later, in the summer of 1885 he died, aged 44 years, from injuries received in an accidental fall. He is buried at Sparkill, New York, under a monument shaped like an obelisk.*

The steamship *Dessoug* in its berth at Alexandria has the encased obelisk slid thru an opening in its bow. The shaft was then centered and shored, and the hole closed and made sea-worthy.

move. Gorringe had planned the operation down to the last detail, had had special structural steel footings pre-fabricated and all the required tackle, hydraulic jacks and steel rails shipped to the site. There existed no precedent in moving dead loads amidship larger than 100 tons and this maximum consisted of guns shipped from England to Italy that required the construction of costly and special coastal hydraulic cranes. Gorringe, therefore, had to design and fabricate the moving equipment based on shifting the dead weight of a fragile stone weighing more than twice that maximum.

A pit displacing 1730 yards of earth was first dug about the shaft exposing some twelve feet that had been buried by the accumulated debris. He then dug down to below the footing that the Roman engineers had erected after the transportation operation from Heliopolis. He sheathed the stone with timber to protect the delicate inscriptions and this packing was fastened into place by iron bands which served both for protection and for the fastening of control ropes during the turning operation. Altho Gorringe had calculated that the mid-section stresses were adequate to hold up the end-sections when the shaft was in a horizontal position and balanced on its trunnions, still in order to be safe with a stone that had been subjected to 3000 years of exposure, he applied a two-cable suspension truss that lifted about 30 tons from each end section. Timber shears then lifted the obelisk weight and transferred it to the trunnion pillow-blocks. These in turn rested

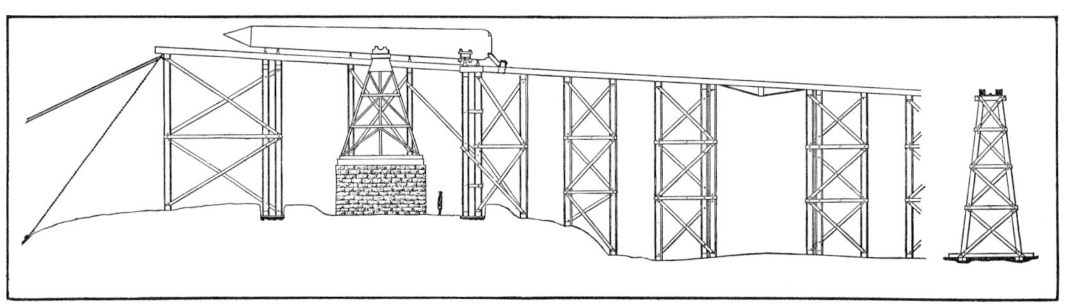

The route of the obelisk from its Hudson River stage to its final position in Central Park led over roads and trestles. Here it terminated over its reconstructed pedestal. Trunnions will enable it to be swung vertically and then lowered.

on steel framework footings that were placed on either side of the obelisk and rested on two masonry piers, built on the sides of the obelisk pedestal.

Tackle was provided to control the turning of the obelisk to the horizontal position but when this operation brought the shaft to a slant of about 45° the controlling rope broke and the purposely loaded front end swung down rapidly and partly crushed the staging prepared to receive it. No damage was otherwise done. The obelisk was then lowered into a wooden caisson in order to float it around the breakwater in the shallow bay to the waiting steamship *Dessoug* which had been bought and fitted out especially for this duty. The caisson was pulled thru the water for a circular distance of about 10 miles to bring it to a point only about one mile (across the city) from its former position, for Gorringe could not get permission to take his burden thru the city. This forced manoeuver raised the costs by $21,000.

The obelisk was lowered by hydraulic jacks for a distance of 43 feet down two stacks of timber in gradual steps so as to bring it to the level whereby it could be warped onto the level of its caisson bed. The main pedestal stone and the lower stones forming the footing as assembled by the Romans were then removed, marked and boxed, and transferred to the *Dessoug*.

When the obelisk was encased in the caisson it was pushed out by hydraulic jacks into the pounding surf of the Mediterranean that threatened to wreck the caisson or break the obelisk. Finally, on March 31, 1880, the caisson was water-borne and a tug pulled it for the ten-mile course around the jetty to the dry-dock where

COURTESY N. Y. PUBLIC LIBRARY

With the trestle removed, the obelisk is swung into a vertical position over its pedestal. The steel cable truss lifted the end loads of the fragile stone; the footing rods kept it from slipping.

the steamer *Dessoug* was supported. The obelisk, still sheathed in its wooden protective cover, was slid thru a hole opened in the forward starboard end of the vessel, until it rested, point forward, in the ship's hold. It required a combination of a crane on the arsenal quay and a chartered floating derrick to lift the 50-ton pedestal, then the base stones, and to lower them into the hold of the steamer.

Neither crane nor derrick had the capacity for the pedestal stone alone, but together they were able to lift it from the lighter deck and lower it into the aft hatchway of the *Dessoug*. The obelisk was bedded down in soft pine for the trans-Atlantic trip and shoring to the ship's timbers was installed to prevent slide or motion during the journey. The *Dessoug* sailed on June 12, 1880, just eight months after Commander Gorringe had first arrived in Alexandria.

On arrival in New York on July 20, 1880 the ship docked at West 51st Street and cranes unloaded the pedestal stone. This was then slung on chains under a four-wheel carriage drawn by 32 horses. Hydraulic jacks had to push the carriage out of occasional ruts made in the road by the heavy rear wheels. A new site on a knoll was picked near the Metropolitan Museum of Art to receive the monument. From this knoll the earth was stripped and the granite rock was leveled. On this the foundation stones were reset in the exact arrangement and position, and in the same orientation as they had at Alexandria. The iron and steel clamps* that the Romans had used to bind the stone together were replaced by modern counterparts. Into the interstices between the foundation stones formerly left empty by the Romans, Gorringe placed lead boxes containing state documents, records and contemporary data relating to the obelisk, coins and medals of the United States, the Bible, works of Shakespeare, a dictionary, nautical tables and models of various tools then in common use. The remaining voids were filled with cement. The obelisk itself was made ready for disembarking, which proved no easy matter.

Just as the functionaries at Alexandria had made it difficult for this ancient stone to leave, so, upon its arrival, were some as unfriendly in greeting it at the new shores. The dry-dock owners, demanding exorbitant rates, caused the *Dessoug* to be moved to Staten Island where her bow was lifted and supported. The hole in the starboard bow into which the obelisk had been admitted at Alexandria was now re-opened, the obelisk was raised and turned and then moved out onto a wooden landing stage built on piles. The withdrawal was made upon steel cannon-balls moving in channels.* This method was used until it was found that the great pressure caused the splitting of the channel web. The

**Two metallurgical notes developed as a result of this project. One was the discovery of the iron clamps binding the pedestal stones together. These were fashioned by the Romans and were covered with lead to keep them from oxidizing. They were 1894 years old when a piece of this iron was subjected to chemical analysis by Dr. Wendel of the Albany Steel Works who, in his report, stated that a clean fracture revealed a highly carbonized and granular but tough-looking metal similar to that of puddled steel. It showed, further, that it was composed of iron and 0.5% carbon giving the metal a hardness of rail steel, low silicon and phosphorus content, showing a sound method of manufacture, and 0.218% calcium, showing a plentiful supply of lime used as a flux. The small amount of slag as well as fine fractures indicated frequent re-working. The lead encasing the iron was also found of high standard of purity, 98.9% Pb, 0.06% Cu, 0.052% Ag. (Report by Prof. R. H. Richards in* GORRINGE, *page 172). A specimen of the bronze forming the crabs that supported the four corners of the obelisk was submitted by Commander Gorringe to Prof. F. A. Genth for analysis. The results showed—copper 90.700%, tin 8.127%, lead 0.312%, iron 0.201%, cobalt 0.108% and traces of nickel and sulphur.*

**Count Carburi de Ceffalonie (See Bibliography) had moved the pedestal stone that was to form the base for the equestrian statue of Peter the Great from the Forest of Karelia to St. Petersburg in the 1770s. This stone weighed 600 tons. Two hundred men moved this mass 600 feet a day over a hill, across a marsh and a river by the use of pulleys and capstans for a distance of eight miles. He used iron cannon-balls moving in iron channels to do this. The same method was used to move the New York obelisk into the* DESSOUG'S *hold.*

cannon-balls were therefore replaced by rollers moving over flat steel bars, similar to those used on a marine railway.

Wooden pontoons carried the obelisk to a slip on the West 96th Street pier in the Hudson. From here the obelisk was pulled by block and tackle over prepared ways and bridging, crossing the Hudson River Railway, down thru streets and avenues and over the undulating Central Park terrain for a total distance of two miles. In this instance the shaft was gradually elevated 147 feet from the landing stage elevation to that of fitting into its new trunnion position; at one part of its route the elevation reached its high point of 230 feet above its starting elevation.

Obelisk, pedestal and donkey-engine were fixed to a bed moving on rollers over timber ways. The engine drum reeled up the rope passing thru blocks attached to a distant anchor, thus drawing the bed and load forward. In this way the timbers over which the carriage passed were freed, moved forward and reset on level crib-work to be used over and over again; the anchor also was moved forward. It required 112 days for the obelisk to be moved from the river to the final site. There, on January 15, 1881, the hydraulic pumps lowered the obelisk with its trunnions into the pillow-blocks and it was then swung into position over the relocated pedestal.

Whereas the ancient Egyptians placed their obelisks directly on their pedestals, the Romans reset the shafts on bronze astragals in a space between pedestal and shaft to provide space for their hoisting and lowering equipment. Gorringe combined both methods by setting the base of the obelisk directly upon its pedestal and yet restoring its four bronze crabs similar to the two which remained where the Romans had installed them. However, replicas of these crabs, weighing 922 pounds each, were substituted by Commander Gorringe by castings made at the Brooklyn Navy Yard at his own expense. The original two are now on view within the nearby museum walls. Upon the claws of the crabs were engraved the main points of the history of the obelisk, thereby repeating what the Romans had done in placing Latin and Greek inscriptions upon the claws of their astragals. The text upon the original claws states that "in the eighteenth year of Augustus Caesar when Barbarus was prefect of Egypt this was placed [here]. Pontius [was] architect."

With due ceremonies and before an assembly of 10,000 people gathered in the bitter cold of January 22, 1881 the obelisk was lowered and settled into its present position. It represents a triumph of intelligent and devoted engineering by Commander Gorringe and it is the oldest monument of that size in the New World. In its proper setting on the restored 11½ foot pedestal this ancient shaft now towers 81 feet over a new landscape.

The rough and worn condition of the westerly face as compared to the three other faces has been the cause of considerable speculation as to how this uneven surface came about. It seems most probable that in the destructive rage of Cambyses, King of Persia (525 B.C.), the fires of burning Heliopolis caused the surface to chip and disintegrate. Both pharaohs honored by this monument had, in their days, conquered Persia. It therefore seems that in order to efface the record of the conquerors, fires built around these monuments would have caused this damage. Similar results were evident in the burnt-out stone buildings in the recent war.

Shortly after the erection of the obelisk it was noted that flaking of the surface had become serious. It was therefore examined by a commission who made recommendations for its preservation. After two and a half barrels of flakes and chips had been removed by hand, a water-proofing process consisting of warming the granite surface and brushing on a paraffin solution, was carried out in 1885.

FROM ILLUSTRATED LONDON NEWS, OCT. 27, 1877

The pontoon ship *Cleopatra* containing the obelisk, on her beam ends and abandoned by the towing ship *Olga*, in the Bay of Biscay. Six seamen, sent to the right boat, were lost.

Mr. Vanderbilt was so impressed with the success of the removal and erection of the obelisk that he increased his bid of $75,000 to the actual cost of $104,000.

The London Obelisk

THE HISTORY of this monument is like that of the one in New York in that both were erected by Thothmes III (18th Dynasty, 1461 B.C.). Both stood as a pair before the temple of Tum in Heliopolis and were then moved to Alexandria during the reign of the Roman conqueror, Augustus Caesar. Here they stood at the entrance of a water inlet into the Roman show-place, the Caesareum. It was probably during one of the earthquakes in 1301 or 1303 that this one of the pair was thrown to the ground. Here it gradually became buried in the sand.

At the battle of Alexandria in 1801 between the French and the British forces it was decided by the officers of the British Army to take with them a trophy to commemorate their victory, and the prone monument suggested itself as an appropriate item. A sunken French frigate was raised and it was intended that she should bear the obelisk, which was to be drawn down a pier into the water for embarkation. A storm arose and washed the ship away, and the plan was abandoned. In 1819 there was a formal presentation of the monument to the British but the plans for moving the obelisk faded with the passage of time. There was a revival of interest in 1832 but it was not until 1867 that Lt.-General Sir James Alexander began the task of removal in earnest. From him interest in the enterprise shifted to Prof. Erasmus Wilson, F.R.S., who commissioned the civil engineer John Dixon,* on January 30, 1877, to prepare plans for embarking, transporting and erecting the monolith at the fixed cost of $50,000.

Nearly 2000 years had passed since similar projects had been undertaken but of this, the removal of the obelisks by the Romans, no records existed. Forty years earlier the French had moved an obelisk from Luxor to Paris using manpower and tackle. The removal of the obe-

*John Dixon (1835-1891) came from a Newcastle family of engineers and had an active career in the construction of railways, bridges, ports and wharves in many corners of the world including England, Spain, Brazil, Portugal, Egypt and Gibraltar. He built the first railroad in China, from Shanghai to Woosung, in 1875, a line permitted to operate for two years before it was torn up by the mandarins as an evil influence. Sir Erasmus Wilson chose Dixon for the task of moving the obelisk from Alexandria, an enterprise that proved both hazardous and very costly to the engineer who, in addition to the technical requirements of the operation, was called upon to suffer heavy financial risks.

FROM ILLUSTRATED LONDON NEWS, MARCH 10, 1887

The removal of the London obelisk with its uncovering, top left. Shown also are details of the pontoon boat *Cleopatra* and, in the center, boat and towing ship prior to the coming of the tragic storm.

lisk to London coincided with the development of steam power and therefore required the use of tools of greater power and more recent development than those which the Romans or French had used, for Mr. Dixon thought in more modern terms. The enterprise was the first to be privately financed (as was the New York venture two years later) and with it came the further innovation of an iron vessel, steam power, hydraulic jacks and modern mechanical tools. A steam-propelled vessel towed a steel caisson containing the ancient stone.

The shore near which the great stone lay buried was shallow and encumbered with sunken walls and shoals and was subject to frequent gales. A special vessel was therefore required. Mr. Dixon decided on the construction of an iron cylinder to enclose the obelisk as it lay on the shore, then (like a parabuckle) to roll the assembly into the sea and finally to have it towed to England. This large cylinder-like ship, the *Cleopatra*, was built on the Thames in sections and was similarly shipped to Alexandria for a bolted assembly. The cylinder was 93 feet long and 15 feet in diameter. Nine watertight bulkheads divided her into ten compartments. The inner chamber was lined with elastic timber cushions and every precaution was taken to minimize the stresses on the stone during the lowering down to the water, then against the pitching and the rolling of the sea.

The area around the stone was cleared and excavated to the level of the lower face. Then stout timbers were placed under it and with hydraulic jacks it was moved so as to be parallel with the shore line. The diaphragms surrounding the obelisk were next put on and piece by piece each part was riveted and calked to the preceding member. The sea-wall was then demolished and with tools and dynamite stones were blasted and cleared and the approach to the sea was graded. Fearing unforeseen impediments it was deemed necessary to strap two rings of timber nine inches thick to both ends of the cylinder thereby forming the equivalent of two wheels 12 feet long and 16½ feet in diameter. Wire hawsers from moored lighters made nine turns about the caisson and this provided the downward pulling force. To avoid jumping and shock, the pull was checked by hawsers controlled on the shore side.

On August 28, 1877 the hawsers on the lighters were drawn in and the large cylinder began to roll. The next day, nearly in floating position, it was discovered that a sharp stone had pierced the skin of the caisson, and that someone had forgotten to close the bulkhead doors. The cylinder had therefore half-filled with water. Repairs were made and the vessel was pumped dry. She floated, was taken to dry-dock and bilge-keels 40 feet long were riveted on. A cabin and bridge were added topside, a mast was stepped and a rudder hung; then ballast was added. With full gear the *Cleopatra* displaced 290 tons. On September 21, 1877 she was towed out of Alexandria by the steamer *Olga*, and at sea a pitching of 16 per minute was noted. Stops were made at Algiers and Gibraltar. On October 14th a storm arose and, not too seaworthy at best, the cylindrical vessel yawed considerably until a great wash shifted the 20 tons of iron ballast causing the vessel to go over on her beam. In this difficulty the captain of the *Olga* called for six volunteers to get to the *Cleopatra* to help tighten the ballast. Six seamen manned their boat but on nearing the caisson a heavy sea rolled over them and they were lost. Later a line was reached from the *Olga* to the *Cleopatra* and Captain Henry Carter of the *Cleopatra* and her crew were taken aboard. The *Cleopatra* was then lost from view and was deemed as having foundered. The *Olga* there-

FROM ILLUSTRATED LONDON NEWS, SEPT. 21, 1878

The obelisk is placed on its pedestal on the Thames Embankment. The lower view shows it raised within its timber tower while held by a long steel collar. It is turned on pivots attached to this collar (top, right) and in the top view is shown resting on the pedestal.

upon headed for England. Later the caisson was sighted by another ship and was towed to port.* On January 20, 1878, she was drawn up the Thames estuary and taken to her final berth.

A very sound concrete base was laid on the packed clay of the Thames Embankment and preparations were made for the raising of the shaft. At high tide the *Cleopatra* was grounded, on a sunken timber cradle. The pieces comprising the ship were then removed, part by part, the great stone was raised by hydraulic jacks, and with screw traversers it was slid along the embankment until its center of gravity rested exactly over the prepared pedestal. This pedestal was cut for the occasion, the one at Alexandria having been left to remain there.

Since this monolith was found horizontal and had remained so ever since, a problem was faced in turning it upright. This was done by erecting a huge timber structure, the main upright of which consisted of four elements, each element in turn consisting of six timbers over 60 feet long and 1 foot square. These units were braced by shores and tie-beams. Two horizontal girders were then constructed and they rested on wooden blocks fitted in between the main uprights. As a next step there was fitted about the wood-sheathed obelisk a wrought-iron jacket 20 feet long having on opposite sides knife-

The caisson and cargo were claimed as salvage and a prize which GORRINGE reports (page 105) as either £2,000 or £7,000 was settled upon.

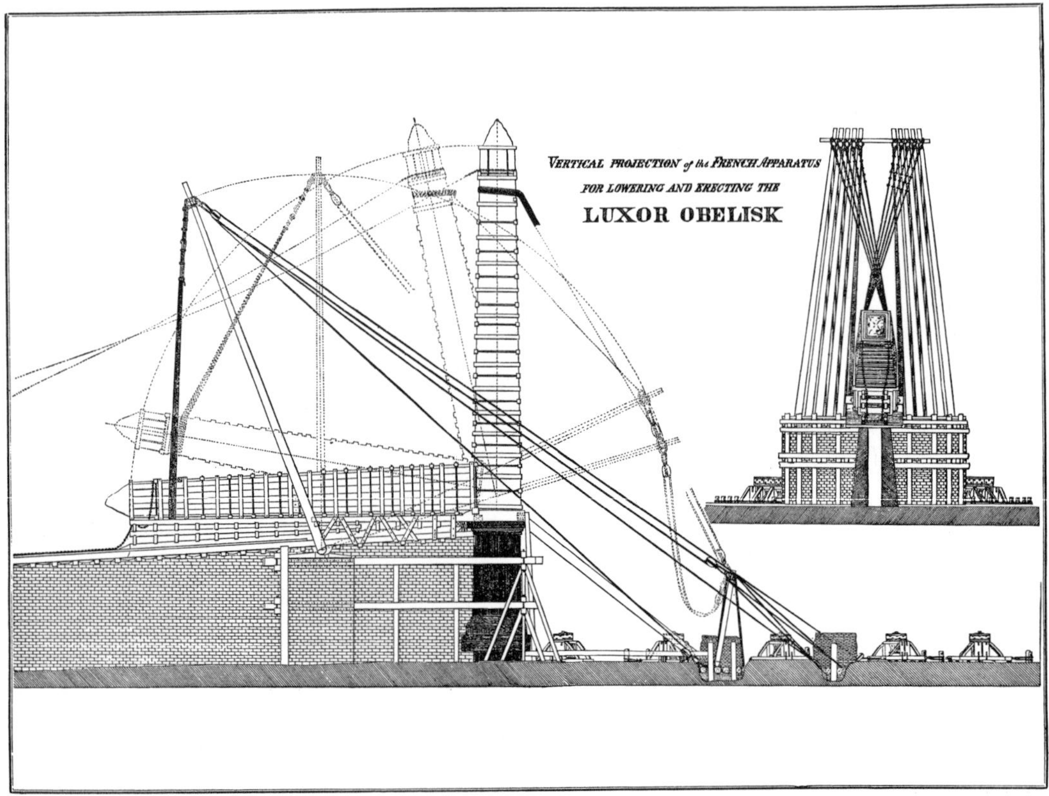

FROM GORRINGE, EGYPTIAN OBELISKS, 1882

The group of masts (see also view at right) used by Le Bas to lower the Luxor obelisk intended for Paris is shown at the left. With the obelisk vertical, the mast tops are projected to the right, but as it was tipped to the left, the masts followed and took the stone's weight until it rested horizontally.

edge pivots at the exact center of gravity of the obelisk, one pivot each resting upon boxed girders. A stirrup strap passing around the foot of the obelisk was bolted to the jacket; this kept the shaft from slipping downward during the turning operation. The movable horizontal girders were designed to be of sufficient strength to carry the full load of the obelisk after they had both been raised and the monolith swung from the horizontal to the vertical position. The raising of girder and obelisk was by means of hydraulic jacks placed under the girder ends, and as the girders rose, timber blocking was placed under them.

When the obelisk had been sufficiently raised, the new pedestal consisting of three steps and main section composed of five courses of masonry having a total height of 18 feet 8 inches was erected under the computed (and actually determined) center of gravity. Into a section of this built-up pedestal were deposited two earthen jars containing various coins and common objects, sacred and profane writings, technical standards, almanacs and directories. On September 12, 1880 the shaft was turned and on the next day it was lowered onto its pedestal.

Three bronze supporting crabs were cast and added to the one original crab and these were inserted at the corners of the obelisk to reinforce its rounded base. Bronze sphinxes, copied after Egyptian models, were added to the area. On the pedestal were later inscribed the names of personages and dates connected with the origin and transportation of the obelisk and among these were listed the names of the six seamen lost in their efforts to retrieve the storm-bound *Cleopatra*.

Mr. John Dixon, C.E., who so carefully computed the mechanical and technical details of the enterprise also acted as the contractor and did not do at all well in his finances. Since the original guaranteed amount of $50,000 was exceeded, the sum grew, because of unexpected losses and changes, to a figure exceeding twice the original amount. Mr. Dixon must, however, have found satisfaction in a task well done, for he received no other formal recognition.

For preserving the face of the obelisk from the dangers of industrial London it was treated with a solution of sodium silicate (water-glass) and later with a solution of gum dammar containing also wax and mercuric chloride.

The Paris Obelisk

It was a Frenchman, Boussard, who found the key to the history of ancient Egypt and her monuments and it was another Frenchman, Champollion, who applied that key to the interpretation of the hieroglyphs on these monuments for our better understanding. It is therefore not surprising that France should be the first modern nation to arrange for the acquisition of one of Egypt's obelisks. It would not be France's first such acquisition, for one, 56 feet 9 inches tall, stands in Arles, probably brought there by Constantine. This obelisk was not cut from seyenite, but is of grey granite, and therefore may not be of Egyptian, but of Roman, origin.

Napoleon's activities in the campaign of the Nile may have been the occasion for him to decide to take an obelisk to Paris as a trophy. However, it remained for Louis XVIII to sponsor an actual plan, and the upright one of the shafts at Alexandria was given to France. However, this plan was also abandoned and the same stone was later given to America.

Champollion in his travels in Egypt saw the obelisks at Luxor (Thebes) and wrote that one of these, rather than the Alexandrian shaft, should be procured. In 1830 Baron Taylor received a royal warrant to proceed to Egypt and to make arrangements for its transfer. This was followed by the construction of the immense barge, the *Luxor*, in order to ascend the Nile and transport the obelisk that was to be obtained there.

The requirements made of such a barge was that it should be able to navigate the Nile as far as Thebes, drawing, while loaded with the obelisk, not more than six and a half feet of water, be seaworthy for crossing the Mediterranean and the coastal waters of Spain and France, be able to be towed up the Seine and clear all the city's bridges, and still be strong enough to take the obelisk's weight while beached at Thebes. Such a barge was constructed after plans prepared by Baron Rolland and the 5-keeled, 3-masted barge was built and launched at Toulon in July 1830. Personalities and the government of France changed at this time, but the project went on and the naval constructor, M. Apollinaire Le Bas* was commissioned to take the *Luxor* to Alexandria and up the Nile. At Luxor stood two tall obelisks before the great gateway to the temple, as they had stood for 3200 years. Taking advantage of the rise and fall of the Nile during and after the rains, it was arranged to draw the barge at high water to within 430 yards of the standing shaft. The general plan was that when the water receded the stone was to be drawn onto the barge. At the Nile's rise the following year the barge and its cargo were to be floated and so drawn down the river.

In the lowering operation scaffolding was built around the shaft; it was then measured and found to be 74 feet 11.2 inches tall with a damaged tip on the pyramidion. From its dimensions and the weight of a similar stone it was determined to weigh 246 tons. Two of the opposing faces were found to have slight outward curvatures, a condition found in both of the obelisks standing there. It was therefore not considered accidental but of undetermined significance. From the scaffolding, the faces of the obelisks were sheathed in 5½ inch planking and were banded with wooden and cable bands.

The Luxor obelisks are among the best preserved of all those known. Much of the original polish was still retained and the engravings were sharp and clear, often 2 inches deep. In addition to the engraving on the faces, Le Bas found the cartouche bearing the name of Rameses II on the bottom of the monolith.

As can be seen from the figure on page 54, it was the plan of Le Bas to lower the obelisk onto an earth ramp of gentle slope leading to the barge. To make this possible he proposed a double pivot arrangement so that the frame of the massive weight could be shifted from the pedestal to a group of wooden masts rotating on their lower ends.

He first fitted a round oak log about nine feet long to the edge of the obelisk foot about which the shaft was going to be tipped. This log was fitted into another rounded hollow wooden log that formed a bed or hinge to receive the first log. The purpose of this hinge was to prevent crushing of the stone edge of the obelisk foot in the lowering process since this hinge was to bear all of the obelisk's weight during the lowering operation. To the obelisk was then attached an iron chain to draw the shaft from its vertical position and to begin its rotation to the horizontal track. Once the obelisk was tipped to a slant wherein its center of gravity was outside the hinged edge, this chain-gear was no longer required.

In the opposite direction and also from a

J. B. Apollinaire Le Bas was born in 1797 in France. He studied at l'Ecole Polytechnique and entered the Corps of Marine Engineers in 1818. He was placed in charge of the marine museum of the Louvre, a position he held for sixteen years (1836-52). He died in 1873.

FROM WILKINSON, TOPOGRAPHY OF THEBES, 1835

A sketch made on location at Thebes showing the brace of masts rising as the Luxor obelisk is lowered. The men in the right foreground are snubbing the control ropes about a fixed beam. At the raising of the shaft in Paris ten masts, instead of the eight masts shown here, were used.

position near the top there extended a web of ropes forming the restraining bridle. These ropes were connected with a horizontal timber forming the top of a swinging bent the sides of which consisted of eight masts, four on each side of the obelisk. These masts tapered towards their tops, were fitted into the horizontal timber at their tops, and at their rounded bottoms were pivoted in a hollow timber set at a distance from the first pivot (the foot of the obelisk) of about half the obelisk height. While the obelisk was vertical, the trapezoid of masts pointed outward nearly horizontally, but as the obelisk was lowered these masts rose, stood vertically and then continued in their arc of swing until the obelisk suspended from the masts rested horizontally on its bed. The French seamen handling the free ends of the lowering tackle snubbed the ropes about fixed posts and beams and each man had only about 30 pounds of ultimate strain to contend with.

After a period of intense training in which the primitive natives had to be taught the use of the simplest of tools, and after a siege of cholera in which many Arabs died (some overnight), at daylight of the 24th of October 1831 the signal to commence the lowering operation was given. Examination disclosed that a fissure existed on one side, a crack that the original constructors had tried to mend by the use of dove-tailed wooden dowels, the dust of which still remained. Halfway towards the descent the lower end of the obelisk pushed itself into the earth foundation forcing Le Bas to shift his gear in order to push the shaft up and out of the pit thus created. He then found that the timber cross-work upon which the obelisk rested had been crushed and had to be replaced from the slender stock of wood then at hand. Further, the shaft moved laterally on the ways and had to be righted. Finally, after many days and much damage to the precious supplies of wood, the obelisk finally moved towards the barge. Then came a period of eight months of waiting for high water.

Much of this operation was carried on in intensive heat, the recorded temperature on one day reaching 150°F and staying at that heat for four hours.

To get the obelisk into the barge the front section of the boat had to be cut away and raised on shears until the stone was stowed amidship. This section was then lowered and the seam repaired to make the bow seaworthy. On August 18, 1832, the barge with her precious cargo was afloat and the slow journey down the river began. At the mouth of the Nile the steamer *Sphinx* towed the *Luxor* on her seaward journey.

It was not until August of 1834 that the *Luxor* was properly in place quayside at the Seine, its bow opened and 240 men manned the capstans. Five six-fold blocks were attached to a cradle upon which the obelisk was drawn, and these pulled the cradle up a ramp. Forty-eight men at each of the five capstans supplied the required power.

The original pedestal, beautifully engraved, had been left in its original position at Thebes and a new one, consisting of five pieces of stone of a total weight of 236 tons, was shaped and placed on the Place de la Concorde. To draw the obelisk on its last length of journey it was planned to use a new mechanism, the steam-engine, but to the disappointment of the Parisians, the machine failed in its tests and the more dependable, manned capstans were again put to work. The horizontal shaft finally reached its exact turning position over the newly completed pedestal and the process that lowered it at Luxor was, in essentials, repeated in reverse. Ten masts took the load instead of the original eight, and 480 men worked the capstans. On October 25, 1836, in the presence of 250,000 Parisians, M. Le Bas directed the raising of the obelisk which, with one slight mishap, gently rose to its new position and rested on its pedestal of newly cut Breton rock. Five years had passed since it had been lowered at Thebes. Not only were praises of king and engineer inscribed on the faces of the pedestal but the equipment and operation that made the transposition possible were also engraved. To preserve the granite surface in its new atmosphere, the obelisk was coated with a rubber solution.

In comparing the works and plans of the engineer of today with those of his ancient counterpart we can stop to wonder at the scale of their work. So must we equally admire the confidence of the engineer Dimocrates of Macedon who proposed to his chief Alexander the Great that "he should carve Mount Athos into a statue of a man with a city in one hand, and a basin in the other, which should receive all the waters of the mountain and again discharge them into the sea." (BARBER, *page 120*)

DATA ON THE TWELVE TALLEST OBELISKS

OBELISK NAME	PRESENT LOCATION	HEIGHT in Feet	WEIGHT in Tons	ORIGIN and MOVES	SPONSOR of MOVES	WHEN MOVED	ENGINEER	METHOD of TRANS-PORTATION	HISTORICAL
Lateran	San Giovanni in Laterano, Rome	105′–7″	510	Assuan to Thebes to Alexandria to Rome: Circus Maximus San Giovanni, Laterno	Thothmes III, IV Constantine Constantius Sixtus V	18th dyn., 1500 BC 345 AD 357 AD 1588	D. Fontana (1543-1607)	Tower, pulleys, carriage	Broken into three pieces. Three feet of stone has been cut away.
Karnak (1)	Thebes, Egypt	97′–6″	371	Assuan to Thebes	Queen Hatasu	18th dyn., 1625-1591 BC	Senmut		Cut, engraved, polished in seven months.
Assuan	Assuan, Egypt	95′–11.5″	770	Still at Quarry at Assuan					Cut on three sides, lower side still attached to rock. Abandoned because of crack in its center.
Vatican	St. Peter's, Rome	83′–1.5″	361	Assuan to Heliopolis to Rome: Circus Nero St. Peter's	Meneptah I (?) Caligula Sixtus V	19th dyn., 1322-1302 BC 41 AD 1586	D. Fontana	Tower, pulleys, carriage	Uninscribed by Egyptians. Largest entire obelisk outside Egypt. Operations begun 1585, completed 1586; cost $44,000.
Luxor	Thebes, Egypt	82′	284	Assuan to Thebes	Rameses II	19th dyn., 1388-1322 BC			Of a pair with one now at Paris.
Flaminian	Piazza del Popolo, Rome	78′–6″	263	Assuan to Heliopolis to Rome: Circus Maximus Piazza del Popolo	Seti I Augustus Sixtus V	19th dyn., 1439-1388 BC 20 BC 1589	D. Fontana	Tower, pulleys, carriage	Broken into three parts.
Paris	Place de la Concorde, Paris	74′–11″	246	Assuan to Thebes to Paris	Rameses II	19th dyn., 1388-1322 BC	J. B. A. Le Bas (1797-1873)	Pivoted masts, barge	Ancient seam in side. Pedestal abandoned. Operation begun 1830, completed 1836. Estimated cost $500,000.
Karnak (2)	Thebes, Egypt	71′–7″	173	Assuan to Thebes	Thothmes I	18th dyn., 1646-1625 BC			Stands in Hall of Columns at fourth propylon where its mate lies broken.
Campensis	Monte Citorio, Rome	71′–5″	230	Assuan to Heliopolis to Rome: Campus Martius Monte Citorio	Psametik II Augustus Pius VI	26th dyn., 596-591 BC 20 BC 1792	G. Antinori (1734-1792)		Inscription on base by Pius VI attributes it to Sesostris. Discovered in five pieces, restored and erected in 1792. Most of inscription effaced.
New York	Central Park, New York	69′–6″	224	Assuan to Heliopolis to Alexandria to New York	Thothmes III Augustus W. H. Vanderbilt	18th dyn., 1461 BC 13 BC 1880	Amen-men-ant H. H. Gorringe (1841-1885)	Trunnions on pedestal, shipped in steamer's hold	Inscriptions by Thothmes III, Rameses II and Orsoken I. Operation begun 1879, completed 1881; cost $104,000.
London	Embankment, London	68′–5.5″	209	Assuan to Heliopolis to Alexandria to London	Thothmes III Augustus Erasmus Wilson	18th dyn., 1461 BC 13 BC 1878	Amen-men-ant John Dixon (1835-1891)	In floating caisson to England raised by jacks, turned on pivots	One of pair with that now in New York. At Alexandria it fell, probably during earthquake. Operations begun 1877, completed 1880; estimated cost $200,000.
Heliopolis	Heliopolis, Egypt	67′	136	Assuan to Heliopolis	Usortesen I	12th dyn., 2371-2325 BC			Oldest large obelisk known, last one standing in Heliopolis.

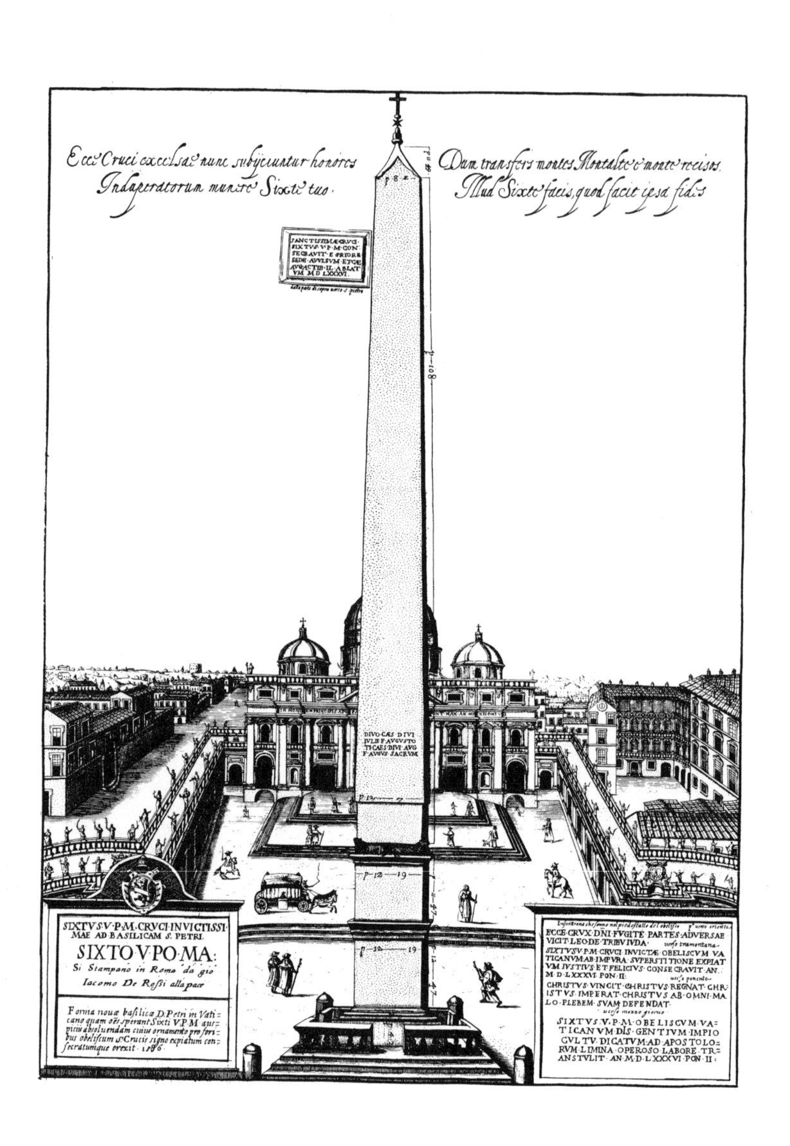
Ecce Cruci excelsae nunc subijciuntur honores
Indaperatorum munere Sixte tuo.
Dum transfers montes Montalte e monte recisos.
Illud Sixte facis, quod facit ipsa fides
SANCTISSIMAE CRUCI
SIXTVS V P M CON
SECRAVIT E PRIORE
SEDE AVVLSVM ET CAE
AVG ACTIB IL ABLAT
VM M D LXXXVI.
DIVO CAES DIVI
IVLII F AVGVSTO
TI CAES DIVI AVG
F AVGVS SACRVM
SIXTVS V P M CRVCI INVICTISSI
MAE AD BASILICAM S. PETRI.
SIXTO V PO MA:
Si Stampano in Roma da gio
Iacomo De Rossi alla pace
Forma nouae basilicae D Petri in Vati=
cano quam oes sperant Sixti V P M aus=
picio absoluendam cuius ornamento pro fori=
bus obeliscum S Crucis signo expiatum con=
secratumque erexit 1586
ECCE CRVX DNI FVGITE PARTES ADVERSAE
VICIT LEO DE TRIBV IVDA
SIXTVS V P M CRVCI INVICTAE OBELISCVM VA
TICANVM AB IMPVRA SVPERSTITIONE EXPIAT
VM IVSTIVS ET FELICIVS CONSECRAVIT AN
M D LXXXVI PON II
CHRISTVS VINCIT CHRISTVS REGNAT CHR
ISTVS IMPERAT CHRISTVS AB OMNI MA
LO PLEBEM SVAM DEFENDAT
SIXTVS V P M OBELISCVM VA=
TICANVM DIS GENTIVM IMPIO
CVLTV DICATVM AD APOSTOLO=
RVM LIMINA OPEROSO LABORE TR=
ANSTVLIT AN M D LXXXVI PON II

BIBLIOGRAPHY

Books relating to Obelisks in the Burndy Library

AGRIPPA, CAMILLO; Trattato di Camillo Agrippa Milanese Di Trasportar la Guglia in su la Piazza di San Pietro, Roma, 1583.

AGRIPPA, CAMILLO; Nuove Inventioni di Camillo Agrippa Milanese. Sopra il modo di Navigare, Roma, 1595.

BANDINIO, ANGELO MARIA; De Obelisco Caesaris Augusti, Roma, 1750-1.

BARBER, F. M., COMMANDER, U.S.N.; The Mechanical Triumphs of the Ancient Egyptians, London, 1900.

DE CEFFALONIE, COMTE MARIN CORBURI; Monument Elevé a la Gloire de Pierre-le-Grand, Paris, 1777.

[ANON.]; Familiaris Quaedam Epistola e Roma in Hispaniam Missa, in Qua Quid Actum sit Die xxix Aprilis, viij Maij, xxvij Septembris in Translatione Obelisci Breviter Explicatur, Roma, 1586.

FONTANA, DOMENICO; Della Trasportatione dell'Obelisco Vaticano, Roma, 1590.
Second edition, Napoli, 1604.

FONTANA, CARLO; Templum Vaticanum, Roma, 1694.

GORRINGE, HENRY H., LT. COMM., U.S.N.; Egyptian Obelisks, New York, 1882. *It forms, with Fontana, the two most important references on the subject.*

JULIEN, ALEXIS A.; Notes of Research on the New York Obelisk, New York, 1893. *Special emphasis on the geological and mineralogical aspects of the subject is here given.*

KIRCHER, ATHANASIUS; Obelisci Aegyptiaci, Roma, 1666. *An attempt to interpret hieroglyphics before the Rosetta key was found.*

[MAGGI, GIOVANNI]; Nova Racolta degl'Obelischi et Colonne Antiche, dell Alma Citta di Roma, Gio. Iac. Rossi, Roma, ca. 1620.

MERCATI, MICHELE, MONSIGNOR; De gli Obelischi di Roma, Roma, 1589.

MERCATI, MICHELE; Considerationi di Monsig. Michele Mercati sopra gli Avvertimenti del Sig. Latino Latini intorno ad Alcune Cose Scritte nel Libro de gli Obelischi di Roma, Roma, 1590.

MEYER, CORNELIO; L'Arte di Restuire a Roma la Traslasciata Navigatione del suo Tevere, Toma, 1685, 1696, 1698. *A number of engineering problems of this competent Dutch engineer including moving and embellishing the Roman obelisks.*

MOLDENKE, CHARLES E.; The New York Obelisk, 2nd ed., Lancaster, 1935.

MUÑOZ, ANTONIO; Domenico Fontana Architetto, Roma & Bellinzona, 1944.

PARSONS, WILLIAM BARCLAY; Engineers and Engineering in the Renaissance, Baltimore, 1939.

VON PASTOR, LUDOVICO; Sisto V il Creatore della Nuova Roma, Roma, 1922.

RODOCANACHI, E.; Les Monuments de Rome après la Chute de l'Empire, Paris, 1914.

SARTON, GEORGE; Agrippa, Fontana and Pigafetta, The Erection of the Vatican Obelisk in 1586, Paris, 1949. *An excellent study of the literature of the subject.*

[SERGARDI, LODOVICO]; Discorso sopra il Nuovo Ornato della Guglia di S. Pietro, Roma, 1723.

WILKINSON, J. G.; Topography of Thebes, and General View of Egypt, London, 1835.

ZABAGLIA, NICCOLA; Castelli, e Ponti, con alcune Ingegnose Pratiche e con la Descrizione del Trasporto dell' Obelisco Vaticano, e di altri, del Domenico Fontana, Roma, 1743. *This sumptuous folio was written for Zabaglia, an ingenious mechanic who could neither read nor write. The story of the transportation of the Vatican obelisk is retold and excellently illustrated with engravings previously cut for Carlo Fontana.*

In addition to the above books, there was obtained, in Rome, in 1950, a volume of considerable importance. This volume is a collection of 11 pamphlets of which nine are in praise of the transportation, erection and consecration of the obelisk, in verse and in prose. All were printed in Rome in 1586-7, thereby indicating the importance and popularity of the event. The remaining two tracts are of meteorological interest. The morocco binding of this volume is of the early 1700s and bears the arms of Pope Clement XI, including repeated imprints of the pontifical device of the stacked peaks and the star, as placed above the obelisks which Fontana moved in the 1580s. Included among these, is the tract by PIGAFETTA, *of which only one other copy in the United States is recorded, that presented to the Library of Congress by the British Museum thru its director, Sir John Forsdyke, in 1945 (see Library of Congress Quarterly, Feb. 1946). The contents of this volume includes:*

DE AGUILAR, IOANNIS BAPTISTAE; In Dedicationem Obelisci Vaticani, Roma, 1586.

[ANON.]; same as in Bibliography, above.

BARGAEIUS, PETRUS ANGELIUS; Commentarius de Obelisco, Roma, 1587.

BLANCUS, GUILIELMUS; Epigrammata in Obeliscum, Roma, 1586.

CATENA, I. HIERONYMUS; De Magno Obelisco Circensi, Roma, 1587.

GACI, COSIMO; Del Trasportamento dell Obelisco del Vaticano, Roma, 1586.

GALESINIUS, P.; Ordo Dedicationis Obelisci, Roma, 1586.

PIGAFETTA, FILIPPO; D'Intorno all'Historia della Guglia, et alla Ragione del Muoverla, Roma, 1586.

[45 POETS]; Sequuntur Carmina a Variis Aucturibus in Obeliscum Conscripta, Roma, 1587.

DOM
ENICO FONTANA
DA MI
LI DIOCESE DI COMO ARCHITE
TTO ec
CONDVTTORE ET
RET